Environmental Earth Sciences

Series Editors

James W. LaMoreaux
P.E. LaMoreaux and Associates, Tuscaloosa, USA

For further volumes:
http://www.springer.com/series/8394

Hermes Farfán González
José Luis Corvea Porras
Irene de Bustamante Gutiérrez
James W. LaMoreaux
Editors

Management of Water Resources in Protected Areas

Springer

Editors
Hermes Farfán González
José Luis Corvea Porras
Parque Nacional Vinales
Centro de Visitantes
Pinar del Rio
Cuba

James W. LaMoreaux
PELA GeoEnvironmental, Inc
Tuscaloosa, Alabama
USA

Irene de Bustamante Gutiérrez
Geology Department
Alcalá University
Madrid
Spain

and

IMDEA Water
Madrid
Spain

ISBN 978-3-642-16329-6 ISBN 978-3-642-16330-2 (eBook)
DOI 10.1007/978-3-642-16330-2
Springer Heidelberg New York Dordrecht London

Library of Congress Control Number: 2013930137

Printed on acid-free paper

Springer is part of Springer Science+Business Media (www.springer.com)

Management of Water Resources in Protected Areas: An Introduction

Biodiversity is one of the most important elements in defining a protected area. However, the protection of these areas requires a holistic understanding, keeping in mind all of the elements of the natural and social environments.

On a global scale, there are more than 100,000 protected land and marine sites. These areas cover almost 19 million square kilometers (nearly 4 % of the global surface). An immense majority of the sites are terrestrial. Among them, 149 are of exceptional natural value in the most varied environmental conditions. The terrestrial aquatic ecosystems only occupy a small area of the planet, but it may consist of the most threatened biomass and habitat by human activities (Dudley 2008).

Natural ecosystems are heavily dependent on water, as it is essential for the development of life. The ecology and landscape play an important role in the quality and availability of water. It is no coincidence that exceptional hydrological phenomena are found in protected areas. Such is the case of the geothermic occurrences (principally, geysers) in US Yellowstone National Park, the oldest park in the world. The Ramsar wetlands (where the ecosystem dependency on water is strongly evident), the Iguazu Falls (on the border of Argentina and Brazil), or the Zapata Swamp (the largest of its kind on the Caribbean islands) further exemplify this point.

The relationship between the protected areas and the conservation of terrestrial waters is complex. Many real and perceived incompatibilities and challenges arise when considering this relationship (Dudley 2008). However, in many cases, the conservation strategies of the hydraulic resources in protected areas are ignored, or simply deprived of the attention they require. There are many types of suitable management strategies for planning and protecting our valuable treasures. Therefore, water resources management in protected areas is an issue not to be separated from the rest of the conservation measures. The relative considerations to terrestrial waters should be integrated in the administration of all the outstanding protected areas that, on the other hand, should be negotiated in function of their bioregional context and of that of their hydrographical basin in the widest sense (Dudley 2008).

The first *Symposium for the Management of Hydraulic Resources in Protected Areas* held in Viñales National Park, western Cuba, was intended to be a framework of communication about experiences with water resources management in protected areas. Advances in research and possible solutions to the problems within these areas were discussed.

Forty papers from Europe and America were accepted in this meeting. They are grouped under six main parts. The first part is dedicated to **Purification and Reuse of Wastewaters in Rural Communities**. Four chapters (10 % of the book) are related to the theoretical aspects of these processes and present several case studies, especially those that refer to extensive methods. These methods comprise multiple environmental compensations as compared with conventional or intensive treatment systems. The most important are low energy consumption, CO_2 absorption, landscape integration, new habitats for flora and fauna, low sludge production, and solid waste reuse (de Armas et al. 2006). These characteristics render these methods suitable for the integration of protected areas. Proposals to integrate these strategies in the normative document and management plan of protected areas and two examples of their application in Spain and Mexico are presented.

The **Impact of Public Use on Water Resources** (8 % of the book) is analyzed in the second part. The first chapter shows a combined strategic environmental assessment and impact characterization procedure to analyze the impact of human activities on water resources in protected areas through an example from Salamanca, Spain. Other two chapters show the effect of human activities in coastal karst aquifers of Cuba by means of hydrogeochemical analysis. One evidences the human influence over the chemical denudation rate, and the second examines it on the acting hydrochemical processes.

For protected areas, groundwater vulnerability maps are highly desired as management tools, because they help in delimiting protection zones, classifies their importance, and thus show where most management attention is required (Williams2008). Several studies (13 %) of **Vulnerability and Risks of Aquifers** are presented in the third part. Research related with vulnerability of groundwater and strategies for defining the protection zones in selected study areas are presented in this part. The part begins with two chapters on the use of geophysical methods in sensitive zones of Cuba and Spain. The next chapter comprises three papers that exemplify the importance of vulnerability mapping and an adequate definition of protection zone in karst terrains. The fourth chapter the results of the application of EPIK and PaPRIKa in tropical karst areas. The part ends with the first approach for the assessment of groundwater resource protection zone at Viñales National Park, western Cuba.

The applications of Geographic Information Systems (GIS), remote sensing, mathematical models, and hydrochemical studies are presented as tools for the **Design and Management of Water Resources in Protected Areas** of Cuba, Ecuador, and Spain, exemplified in seven chapters. In this part (18 %), the use of GIS and remote sensing is presented in three chapters, one of which determines flood risk assessment in a river basin at the southwest of Salamanca, Spain. Other two chapters show the application of a distributed water balance method model for assessment water excess in the high and medium basins in Ecuador and other, the use of GIS as a platform for integrate diverse database and as results, a 3D geomodel design in a detrital aquifer located in

Madrid, Spain, are presented. An application example of a simulation model based on system dynamics in a protected area and its urban surrounding area in Cuba is presented. Two studies in wetlands are also discussed: one, related to hydrogeochemical studies, is used to define the relationship between surface water and groundwater in an important coastal wetland of Almería, Spain, and the second deals with the analytical framework for the study, planning, and management in an important Ramsar site in Ecuador. The management techniques applied to small watersheds in the mountain karst of the Humid Tropics are exemplified in the case of the Santo Tomas Cave hydrologic system, highlighting the most hydrological features controlling surface and underground runoff and the related biogeochemical hydrodynamics.

Forty-three percent of the book is dedicated to exemplifying the importance of **Research and Monitoring of Water Resources in Protected Areas**. A great diversity of studies on protected areas of America and Europe are presented. The importance of fundamental and applied research in these areas and the design and operation of bio- and hydrochemical monitoring networks are expressed in the chapters presented. Major and minor changes in biodiversity and in the physical framework sustaining the local environment could not be detected without efficient operation of properly designed monitoring networks (Molerio and Parise 2009). Adequate research on the environmental and particularly on the ecohydrological variables is still needed in protected areas. In the particular case of the tropics, the transport of nutrients remains one of the most important tasks to be achieved.

This part begins with a preliminary hydrogeological characterization in a protected area of the Dominican Republic, followed by a chapter about underground water of deep circulation in karst terrains of western Cuba and a study of the capacity of recharge in a protected area of Mexico. Two chapters on rainfall-runoff relationships in Spain and Cuba and nine related to studies of chemical and microbiological composition of water showed interesting examples of contamination by bacteriological and emergent contaminants. A study of seawater intrusion by means of hydrochemical models and geophysical methods in coastal karst aquifers show the importance of multidisciplinary studies. The part concludes whit a study about the seasonal behavior of the vegetation at the complex lagoons.

The last part is dedicated to **Information, Popularization, and Training**, occupying 10 % of the book. Four chapters are included, one of which is related to the strategies used in Cuba for effective management of water. The second one is dedicated to exemplifying the use of geodiversity and hydrological heritage as a tool for educational itineraries that help promoting their conservation through knowledge. In the third chapter, recent observations on the geology of the Cuban occident as well as the most important aquifers and the geomorphology of their basins are exposed. The last chapter is a revision of the importance of Alexander von Humboldt's work in Cuba that is integrated nowadays in the National System of Protected Areas in Cuba.

The diversity of topics discussed in this book makes it a valuable consulting document for the personnel that work directly or indirectly in the field of water research, especially in protected areas. Thus, they deal with common interests identified in protected areas and underline the importance of knowing properly the qualitative and quantitative characteristics of water resources, in order to establish efficient measures for management and conservation of natural resources in protected areas.

References

De Armas A, De Cózar JM, Sánchez J, Márquez M, Martel G, Gutiérrez J, Vera L (2006) La Integración Ambiental y Social de los Sistemas de Depuración Natural (SDN). In: Gestión sostenible del agua residual en entornos rurales. Proyecto Depuranat, Instituto Tecnológico de Canarias (ed.), Santa Cruz de Tenerife, Spain

Dudley N (2008) Directrices para la aplicación de las categorías de gestión de áreas protegidas. Gland, Suiza: UICN. x + 96pp

Molerio Leon L, Parise M (2009) Managing environmental problems in Cuban karstic aquifers. Environ Geol 58:275–283

William P (2008) World heritage cave and karst. A thematic study. IUCN-WCPA. World Heritage Studies. No. 2. 57pp

H.F. González, J.L.C. Porras
Viñales National Park, Visitant Center
Km 23 Carretera a Viñales Viñales Pinar del Río, Cuba
I. deBustamante
IMDEA water. Calle Punto Net n° 4. Edificio ZYE 2ª Planta.
Parque Científico Tecnológico de la Universidad de
Alcalá, Alcalá de Henares Madrid, Spain
J.W. LaMoreaux Chairman
PELA GeoEnvironmental, Inc Tuscaloosa, Alabama, USA
L.F. Molerio-Leon
Inversiones Gamma, S.A

Organizing Committee

President

Dr. José Luis Corvea Porras
Director Viñales National Park. Cuba

Executive Secretary

Dr. Juana María Sanz
Project agent. IMDEA water. Spain
Lic. Ileana Fernández Santana
Comercial Specialist. ECOVIDA. Cuba
Lic. Yudalsis Córdova Domínguez
Environmental Specialist. Viñales National Park. Cuba

Scientific Secretary

M.Sc. Hermes Farfán González
Environmental Specialist. Head of Research and Monitoring Program. Viñales National Park. Cuba
M.Sc. Yoel Martínez Maqueira

Head of Management Plan. Viñales National Park. Cuba
M.Sc. Narciso López Rodríguez
Specialist in Environmental Service. ECOVIDA. Cuba

Scientific Committee

M.Sc. Carlos Aldana Vila. National Center of Protected Areas. *(CNAP). Cuba*

Dra. Irene de Bustamante. Deputy *Director, IMDEA water Institute, Consolider TRAGUA. Spain.*

Dr. José Luis Corvea Porras. *Director. Viñales National Park. Cuba*

Dr. José Antonio Díaz Duque. Polytechnic Superior Institute. "*José A. Echeverría". Cuba*

M.Sc. Carlos Díaz Guanche. *Dpto. Geology, Univ. Pinar del Rí. Cuba*

Dr. Francisco J. Elorza. *President of AIH-Spanish Group. Spain.*

Dr. Juan Reynerio Fagundo Castillo. *CENAMENT. Cuba*

M.Sc. Hermes Farfán González. *Environmental Specialist. Head of Research and Monitoring Program. Viñales National Park. Cuba*

M.Sc. Sandra Fernández. *Environmental Specialist. Viñales National Park. Cuba*

Dr. Eloy García Calvo. *Director IMDEA water, Consolider TRAGUA. Spain.*

Dr. José Luis Goy Goy. *Dpto Geology. University of Salamanca, Spain.*

M.Sc. Nelson A. González Cabrera. Research, *ECOVIDA. Cuba*

Dra. Patricia González. *CENAMENT. Cuba*

Dra. Rebeca Hernández. *Dpto. Geology, Univ. Pinar del Río. Cuba*

Dr. Jim LaMoreaux. *Editor in Chief, Environmental Earth Sciences and Carbonates & Evaporites*

Geol. M.Sc. Leslie Molerio León. *GAMMA S.A. Cuba*

Dr. Ernesto Mujica Benítez. *Head of Research. ECOVIDA. Cuba*

Dr. Jesus Pajón. *Institute of Anthropology. Cuba.*

M.Sc. Robert Ramirez. *Dpto. Geology, Univ. Pinar del Río. Cuba*

M.Sc. Aldo L. Ramos. Research, *ECOVIDA. Cuba*

Dr. Iván Relova Delgado. *Center of information management and energy development. Cuba*

M.Sc. Ana Belkis Rodríguez Torres. *Head Environmental Unit, Minister of Sciences, Technologies and Environment, Delegation of Pinar del Río. Cuba*

M.Sc. Matías Román Pérez Duarte. *National Institute of Hydraulic Resource (INRH), Delegation of Pinar del Río. Cuba*

M.Sc. Carlos Rosas Saavedra. *Head of Conservation. Flora and fauna Company. Cuba*

Dr. Juana María Sanz. *Project agent. IMDEA water. Spain*

Acknowledgements

The present book is a result of the effort of a group of enthusiastic specialists in America and Europe. The greater part of the work reported on takes place in protected areas. Contributions were revised by a Scientific Committee consisting of 25 specialists from Cuba and Spain. The editors of the book acknowledge with gratefulness their contributions and unconditional support for the final elaboration of the book.

The first symposium on Management of Water Resources in Protected Areas, Viñales 2010, was organized by the Viñales National Park (ECOVIDA), the National Center of Protected Areas of Cuba and the Madrid Institute of Advanced Studies in Water Technologies (IMDEA water) of Spain, as part of the agreement for collaboration among these institutions. We thank equally those people and institutions that supported the organization of the event and the realization of the book.

This book is a contribution to the improvement of the Management Plans of Protected Areas, offering a more holistic focus to the study of those natural ecosystems that come under the diverse handling categories defined by the UICN.

Contents

Part I
Purification and Reuse of Wastewaters in Rural Communities

Chapter 1
Wastewater Problems in Rural Communities, Their Influence on Sustainable Management in Protected Areas

J.L. Corvea, Y. Martínez, A. Blanco, I. de Bustamante Gutiérrez, and J.M. Sanz

Abstract Managing wastewater treatment is a valuable alternative for water resource depuration and reuse in small rural communities. This is actually a common practice and is being implemented in territorial planning, based on sustainable use of the diverse systems. Despite the advantages and positive experiences obtained by the use of these techniques in small human settlements, its influence as criteria to evaluate the effectiveness of the natural spaces management can be understood by the administration and management team as a healthy indicator of the ecosystems coming from "secondary effects and reactions". This paper presents methodological bases for the union of wastewater management as a key factor for standard procedures for protected areas, taking into account the phases of diagnostic, normative, programmatic and cartography.

Keywords Ecosystems • Wastewater treatment management • Natural depuration systems • Spain

J.L. Corvea (✉) • Y. Martínez
Parque Nacional, Viñales, Centro de Visitantes, Km 23 Carretera a Viñales, 22400 Pinar del Río, Cuba
e-mail: corvea@pnvinales.co.cu; corvea@pnvinales.co.cu

A. Blanco • J.M. Sanz
Department of Geology, Universidad de Alcalá, Alcalá de Henares, Madrid, Spain

I. de Bustamante Gutiérrez
Geology Department, Alcalá University, Edificio de Ciencias, Campus externo, 28871, Alcalá de Henares, Madrid, Spain

IMDEA Water, C/ Punto Net 4, 2º piso, Edificio ZYE, 28805, Alcalá de Henares, Madrid, Spain
e-mail: irene.bustamante@uah.es; irene.bustamante@imdea.org

H. Farfán González et al. (eds.), *Management of Water Resources in Protected Areas*, Environmental Earth Sciences, DOI 10.1007/978-3-642-16330-2_1,

1.1 Introduction

Water as a natural resource is an element of major importance for the physical environment as well as for the individuals who live in it. Water controls a majority of the processes that occur in nature, shapes the landscape, is essential for the survival of living beings, regulates the natural cycle of ecosystems, and interacts with a majority of abiotic elements that intervene in the conservation of protected spaces. As a management element, wastewater generated from any source where human activity exists may produce severe damage to natural systems. Therefore, measures have to be taken to avoid environmental problems. A sound water management procedure allows for the preservation of natural ecosystems in optimal conditions for which natural depuration techniques are of great value.

The proposition presented by this paper does not pretend to modify the actual standard methods, but attempts to complement it by adding new procedures to the process of decision making regarding wastewater management.

1.2 Environmental Advantages of Natural Depuration Systems (NDS) Over the Conventional Techniques of Waste Water Treatment

Natural Depuration Systems (NDS) comprise multiple environmental benefits compared to conventional or intensive treatment systems, especially those used for wastewater treatment in rural communities that are closer to naturally protected territories. Among those benefits, the most important are (de Armas et al. 2006):

- Low energy consumption
- CO_2 absorption
- Landscape integration
- New habitats for flora and fauna
- Low sludge production
- Capacity of use products and by-products of the depuration processes like wood, ornamental or livestock feeding species for their later sale (Salas et al. 2006)
- Some other advantages are considered such as the potentiality for environmental education. These facilities are often supported by many people due to the environmental services associated with its function.

According to Zavala (2006), water resources, soils, geology and geomorphology are features that were not incorporated into the guidelines to select sites for conservation, nor integrated into the Management Programmes of the Protected Areas. Therefore, it is necessary to include the abiotic elements as key factors for the management of protected spaces. The use of soils and water should be regulated to avoid negative effects on the biodiversity of the zone.

Wastewaters are usually the cause of severe sanitation problems in rural communities due to the economic constraints that limit the application of appropriate measures (Nogueira et al. 2006). NDS also encompass environmental disadvantages, such as the eutrophication of water reservoirs, producing a spectacular bloom of aquatic vegetation, when the design for the NDS is not appropriated or the terms of operation are violated. Usually excess water discharged into the system and the overload in terms of pollutant concentration is the limiting factors that cause system breakdowns. Another problem regarding the use of NDS is the land extension needed to implement the project.

Assuming Salas et al.'s (2006) statements, it can be summarized that NDS is an appropriate technique that can be extended to protected areas due to its economical, environmental and social benefits. From the environmental point of view, this wastewater treatment technique provides, along with water quality improvement, habitat conservation and perfect integration with the local environment. Regarding economic terms, it reduces costs of implementation and management. Socially speaking, the fact that constitutes a source of local employment and the facility may be of importance as a centre for environmental education for local inhabitants.

1.3 NDS Integration with the Management of Protected Spaces

As appointed by Salas et al. (2006), NDS provide services for decentralized management of wastewater, which are possible sources of other resources for the rural environment if included in the production cycle as an alternative for agricultural activities by turning wastes into agro fertilizers and other purposes.

Most of the protected spaces, no matter its management category, include certain regulations that lead to the conservation of the main natural resource. Those spaces with more strict regulations, such as public use, the presence of human settlements and the existence of economic and social activities, are almost absent. Some others include recreational possibilities as well as the development of activities related to the sustainable management on the local environment; nevertheless, it is less probable that both spaces are not affected by the negative impacts of wastewater flowing out to the neighboring communities.

Yet NDS, as an alternative for wastewater treatment, offers interesting environmental services; the management of the existing capacities in rural settlements associated with protected areas is usually a local or state government or a competition of private agencies.

Although there is proper operation of these technologies, only in fewer occasions the administrations in charge of protected ecosystems make use of them in their management plans. In this sense, the water resource should be evaluated not only

Table 1.1 Proposal for the diagnostic stage

Methodology for the management plan compilation (Cuba)	Proposal for NDS use
I. Diagnostic stage	
1. Description of natural conditions	
Brief description of the territorial context	Identification of water and waste water problems. Types of treatments. Operation conditions. Plans for new facilities for waste water treatment
2. Natural resources	
Geology	Description of the main tectonic features that may influence water dynamic
Climate	Bioclimatic features
Hydrology and oceanography	Hydrogeological features. Rock properties, types and distribution of the aquifer, water recharge and discharge, irrigation and flooding areas, pollution vulnerability
Soils	Hydrological properties of soils permeability, porosity, water retention coefficient and soil depth
Biodiversity	Aquatic flora. Water and water dependent fauna
Landscape variety	Description of aquatic habitats and there conservation
3. Socio-economic features of the area and surrounding	
Description of the economical bases	Water supply system dedicated to industry, agriculture, livestock and other services. Waste water discharge
4. Selection of conservation target	
Significance	Characteristics of water resource directly related to conservation objectives
Peculiarity	Characteristics of water resource as unique conservation object
Threat level	Evaluation of the threats affecting the quality of water resource
Conservation characteristics	Water significance as object of conservation
5. Water problems identification	
Environmental problems	Water related problems
Management capacity of the area	Identification of management capacity of the area
Economic and social issues	Water related problems: quality, supply, treatment
Analysis of research needs to support planning and management	Determine needs for information regarding water, water treatment and reuse
Synthesis of the problem	Description of current water problems as a key element of conservation

because of its natural value, but also to keep it from contamination to maintain ecological integrity.

Table 1.2 Proposal for the normative stage

Methodology for the management plan compilation (Cuba)	Proposal for NDS use
II. Normative stage	
Limits and category of the protected areas	Physical and socio-economical issues related to water problems
Objectives of the management plan	Set objectives with the aim of water resource management
Zoning and regulations for the use of the area	Zoning and regulating water resource
Conservation zone	No intervention on water resources admitted
Public zone	Obligation of using natural depuration systems compatible with the management category
Cultural and historical zone	Enhancement and promotion of the traditional usages of water resources
Zone for genetic resources	Prohibition of human intervention
Restoration zone	Regulation of water pollutants
Administrative zone	Meet the water standards
Socio-economic zone	Regulate the use of chemical and organic substances
Buffer zone	Keep close control on pollution sources
Marine zone	Meet the water standards

1.4 Proposal to Incorporate NDS to the Management Plans for Protected Spaces

Based on the methodology to compile the Management Plans for Protected Areas in Cuba (Gerhartz et al. 2008), and considering the priority attention to water resource conservation, some methodological bases are proposed to include wastewater management as a key element for the protected areas' official documentation.

These bases are not meant to replace any normative document that actually regulates the management of protected spaces. The intention is to include at every step of the current methodology the wastewater issue and the existing variety of techniques for wastewater treatment, without modifying the content and formality of the current procedures.

1.4.1 Some Proposals to Establish the Methodological Bases

- DIAGNOSTIC STAGE: Normally, during this phase the water resource is taken into consideration as a specific section: Hydrology and Oceanography, although it is possible to integrate a group of elements that simplify a detailed description about natural water and wastewater characteristics, attention must be paid to

Table 1.3 Proposal for the programmatic stage

Methodology for the management plan compilation (Cuba)	Proposal for NDS use
III. Management plans	
1. Protection programmes	
Surveillance and protection programme	Set surveillance systems for water resource control and installed depuration systems
Fire protection programme	Protection of water supply sources
Plans against disasters	Control over zones with potential risks, danger and vulnerability to water related disasters
2. Resource management programmes	
Forest management plan	Identify actions for water resource conservation
Management plan for species, habitats and ecosystems	Identify vegetal species compatible with NDS
3. Public use programmes	
Tourism programme	Integrate NDS wastewater stations into visits
Programmes for information, education and interpretation	Introduce activities related to water conservation, use, depuration and reuse
4. Programmes for scientific research and monitoring	
Research programme	Carry out research about water quality, ecosystem conservation and vulnerability
Monitoring programme	Set monitoring network
5. Administration programmes	
Administration programme	Create infrastructural, personal and resource capacities
Programme for compilation and actualization of management plans	Water issues have to be permanent and high-priority
Signalization programme	Introduce signals along natural water bodies, water springs and NDS facilities
Training programme	Include water subjects at every training activity
Public relation programme	Set partnerships on water management
Maintenance programme	Define actions for infrastructure maintenance
Investment programme	Declare explicitly the investments to be carried out in relation with water management

their natural values and socio-economic and cultural features of the sites. Using all this information, a varied description of the current water situation in the area is made, including its potential use and the environmental problems concerning water management (Table 1.1).

- NORMATIVE STAGE: It is important at this stage, to integrate all possible aspects to fulfill current regulations in the country, especially those related to protected spaces, as wastewater is a key element for the management of protected areas. Regulation for the use and actions regarding water resources must be specifically addressed to provide a safe and stable water resource. A sound area zoning must be performed to include any harmful factors (Table 1.2).

- PROGRAMMATIC STAGE: Once this stage is concluded, the ending point of the management plan, it is important to consider the potentialities of each programme in order to integrate activities and actions towards a sound water resource management, whether it is for conservation purposes as a natural element or for later reuse as is shown in the next table (Table 1.3).
- CARTOGRAPHY STAGE: It is recommended at this stage to draw maps representing the diversity of elements related to water resource, such as Slopes, Geomorphological units, Hydrogeology, Aquifer vulnerability, Water pollution sources, Pollution risks, NDS location, and Groundwater protection zones.

1.5 Conclusions

The methodology proposed in this paper might constitute a useful tool for decision makers for the integration of Natural Depuration Systems (NDS), within the Management Plans. Wastewaters are the possible cause of the collapse at any average ecosystem, especially those in rural and conservation areas.

Although NDS are managed by foreign institutions, private or public, data of the results derived from its operations must be part of the management and planning in protected areas.

This proposition will be more complex in those places where a karstic component of the substrate occurs, due to the complexity of the surface and groundwater karst systems.

Acknowledgments Project REAGUAM CGL2009-13168-C03-01 (Ministry of Science and Innovation, Spain), IMDEA-water Foundation (Madrid, Spain), and the General of Research Center and Environmental Services (ECOVIDA, Cuba)

References

de Armas A, de Cózar JM, Sánchez J et al (2006) La Integración Ambiental y Social de los SDN. Gestión sostenible del agua residual en entornos rurales. Proyecto Depuranat: 187–227

Gerhartz JL, Estrada R, Hernández E et al (2008) Metodología para la Elaboración de los Planes de Manejo de las Áreas Protegidas de Cuba. Editorial Félix Varela

Nogueira R, Machado P, Urbaño L, et al (2006) Evaluación Económica y Ambiental de los SDN. Gestión sostenible del agua residual en entornos rurales. Proyecto Depuranat: 230–257

Salas JJ, Sardón N, Martel G et al (2006) Fundamentos de los Sistemas de Depuración Natural (SDN). Tecnologías disponibles. Gestión sostenible del agua residual en entornos rurales. Proyecto Depuranat: 27–53

Zavala Carrión Z (2006) Monumento Natural Marcahuasi, Proyecto de Espacio Natural con Características de Geoconservación. XIII Congreso Peruano de Geología. Resúmenes Extendidos. Sociedad Geológica del Perú: 67–70

It must be highlighted that in the low efficiencies at the beginning of the process, the insufficient cleaning made a direct impact in the failure to eliminate the organic matter, as much of vegetable origin as animal, including the phytoplankton and the existing zooplankton.

As for the cloudiness, the fact was considered of reaching extremely low values for a raw water of superficial origin, which makes the removal quite low, that is to say, only 16.7 % as much for the slow filter as for the plant.

The removal of total and fecal coliforms reached values of 84.3 % and 59.1 % compared with the ones obtained in the waterworks plant, being also higher than the ones obtained in the maturing period.

It must also be underlined that the efficiencies reached in the plant are associated not only with the quick filter installed, but to the disinfection with chlorines that is verified in the place as, at the beginning of the experiment, there were very low efficiencies in the plant due to failures in the disinfection of the water.

4.3.2 *Potentialities of Use of the Slow Filter of Sand*

Keeping in mind an expense $Q = 0.87$ l/min and a supply for rural areas of 40 lppd, as well as operating the filter in a continuous way in two 8-h shifts, by the beneficiary neighbors, it is possible to calculate the quantity of houses and/or inhabitants that will be favored with its use.

The number of houses served by the filter is 4.2, each one with 5 inhabitants; this makes a total of 21 inhabitants.

In a small rural town – 40 houses (200 inhabitants), a battery of 10 filters, makes it possible to guarantee an appropriate service of drinking water. Additionally, for this water to be considered safe from the microbiological point of view, it would be necessary to carry out the disinfection at home level, using solutions of chlorine, or a contact tank could be adapted after the filters where this disinfection is guaranteed.

4.4 Conclusions

1. The use of the slow filters of sand can be a feasible alternative to apply fundamentally in the rural areas that do not have a community supply of water, to purify waters of superficial origin of low cloudiness.
2. Although the raw water used as a source in the research can be considered of excellent quality, according to the results, and due to its origin and period in which the sampling was carried out, it is logical to expect significant variations in its quality.
3. The insufficient cleaning and disinfection of the sand used influenced the extension of the maturing period of the filter and in the removal efficiency

obtained at the beginning of the experiment. In this sense, the maturing time of 20 days can decrease and the efficiencies in color removal and coliforms increase.
4. The low levels of cloudiness of the raw water employed make it impossible to obtain an effluent with more quality.
5. The progressive formation of the ***schmutzdecke*** biological layer, as the filter matured, made the efficiency of color removal, total and fecal coliforms rise respectively to 56.2 %, 84.3 % and 59.1 %; being these acceptable values.
6. The existence of total and fecal coliforms and the high percent that these last represent compared to first ones, make it necessary to establish a disinfection system with chlorine to guarantee safe water.
7. With the slow filter of sand, a speed of filtration of 0.2 m/h and a flow of 0.87 l/min, it is possible to give service to about four families of the rural area, at 40 lppd each.

4.5 Recommendations

1. To propose the use of the slow filters of sand, due to the high efficiency and economy of this kind of filter regarding a conventional purification plant, for groups of isolated houses in rural areas, camps of the Plan Turquino, cooperatives, etc. that do not have a community system of drinking water.
2. To extend the test period of this filter up to the rainy season or with other sources of worse characteristics, as for cloudiness, colour and coliforms, to know its efficiency with more accuracy.

References

AIDIS (1991/1992) La crisis mundial del agua-Conferencia de Río. Ingeniería Sanitaria 45(4)/46 (1, 2): 15
APHA/AWWA/WPCF (1976) Standard methods for the examination of water and wastewater, 15th edn. American Public Health Association, Washington
Buiteman JP (1993) Water treatment 2. Water treatment processes and plants. Part 1. Introduction and design aspects. International Course in Sanitary Engineering. IHE, Delft
Huisman L (1986a) Rapid filtration. Lecture notes International course in sanitary engineering. IHE, Delft
Huisman L (1986b) Slow sand filtration. Lecture notes. International course in sanitary engineering. IHE, Delft
NC 93–02:85 (1985) Higiene Comunal. Agua Potable. Requisitos Sanitarios y Muestreos. Ciudad de la Habana
NC 93-11-86 (1986) Fuentes de abastecimiento de agua. Calidad y Protección Sanitaria. Ciudad de la Habana
OPS (1995) Ponderación de los riesgos microbiológicos contra los riesgos de los subproductos de la desinfección química. Ingeniería Sanitaria 49(1):35–36

Chapter 5
Environmental Impact of Human Activities on Water Resources and Its Characterization for Management and Planning of Natural Areas "Las Batuecas-Sierra Francia" and "Quilamas" (Salamanca, Spain)

A.M. Martínez-Graña, J.L. Goy Goy, I. de Bustamante Gutiérrez, and C. Zazo Cardeña

Abstract This paper presents a combined Strategic Environmental Assessment and Impact Characterisation procedure to analyse the importance and extent of the impact of human activities in natural spaces on water resources. In an initial phase, the different human actions (landfills, etc.) that may cause impacts, whether directly (sewage, etc.) or indirectly (leachate, etc.), are determined and the environmental impact identification map is drafted. Subsequently, thematic and interpretive mapping (Surface Water Quality, Aquifer Vulnerability to Pollution and Vulnerability to Solid Waste) is carried out to assess the effects on water resources. Finally, superimposing the Impact Identification map over the Vulnerability maps (Solid Municipal Waste, Aquifers and Surface Water Quality) provides the Impact Characterisation map, showing the absorption capacity of different sectors to facilitate the setting up of preventive and/or remedial measures by the authorities responsible.

Keywords Environmental assessment • Environmental impact • Water resources • Water quality • Aquifer vulnerability • Solid municipal waste • Salamanca Spain

---

A.M. Martínez-Graña (✉) • J.L. Goy Goy
Department of Geology, External Geodynamics Department, Faculty of Sciences, University of Salamanca, Plaza de la Merced s/n (37 008), Salamanca, Spain
e-mail: amgranna@usal.es; joselgoy@usal.es

I. de Bustamante Gutiérrez
Geology Department, Alcalá University, Edificio de Ciencias, Campus externo, 28871, Alcalá de Henares, Madrid, Spain

IMDEA Water, C/ Punto Net 4, 2º piso, Edificio ZYE, 28805, Alcalá de Henares, Madrid, Spain
e-mail: irene.bustamante@uah.es; irene.bustamante@imdea.org

C. Zazo Cardeña
National Museum of Natural Sciences, Geology Section, C/José Abascal nº 2, Madrid, Spain
e-mail: mcnzc65@mncn.csic.es

H. Farfán González et al. (eds.), *Management of Water Resources in Protected Areas*, Environmental Earth Sciences, DOI 10.1007/978-3-642-16330-2_5,

5.1 Introduction

The impact of human activities on the environment has highlighted the need to regulate these activities in a sustainable manner taking into account the absorption capacity of the natural environment. To this end, a set of environmental tools is established: Strategic Environmental Assessment (SEA) and Environmental Impact Assessment (EIA), to ensure rational use of natural resources, prevention and mitigation of the effects of pollution and natural hazards. This paper proposes a combined procedure using tools from the former procedure (EAE) and Impact Characterisation methods from the latter (EIA). This working method allows the management of a natural space to analyse the environmental impact on water resources of existing activities as well as those anticipated in the pre-project phase. This study was conducted in the natural environment of Las Batuecas-Sierra de Francia and Quilamas, in southern Salamanca province in the Spanish Central System (Fig. 5.1).

5.2 Methodology

The methodology used (Fig. 5.2) consisted of applying a mapping procedure with GIS techniques and quantification of environmental impact on natural factors: water, soil, air and landscape, using EIA methods: evaluation of attributes (persistence, extension, synergy, etc.) and matrices (crossing of actions and factors).

The EIA is carried out in an initial phase by identifying the different human activities (landfills, etc.) likely to cause impacts, whether directly (sewage, etc.) or indirectly (leachate, etc.), mapping the environmental impact and/or technological risks (Fig. 5.3a), whether discrete, linear, zonal or areal.

In the second phase, field sheets or checklists are drawn up where each disposal activity and point is characterised by its intrinsic (geology, geotechnical, surface hydrology, hydrogeology, topography and vegetation) and extrinsic features (type of waste, toxicity, persistence, drainage, covering, size, management, etc.) for a qualitative assessment (High-Middle-Low-Null) of the contamination risks at each site and determining which factors will be affected (water, land, air, landscape, geomorphology, active processes, and socioeconomic elements).

Finally, a series of thematic and interpretive maps was drafted to evaluate the effects of different human activities on the water resources:

1. Surface Water Quality map (Fig. 5.3b), where, on the basis of field surveys, the quality and/or degree of surface water pollution is determined through the Simplified Water Quality Index (SWQI) (de Bustamante 1989), which is a dimensionless number that allows operation with very few analytical parameters while providing guaranteed results (Cubillo 1986). This index is defined by five parameters, and its expression is: $SWQI = T (A + B + C + D)$ where T is the river water temperature measured in °C. A is the chemical demand for oxygen

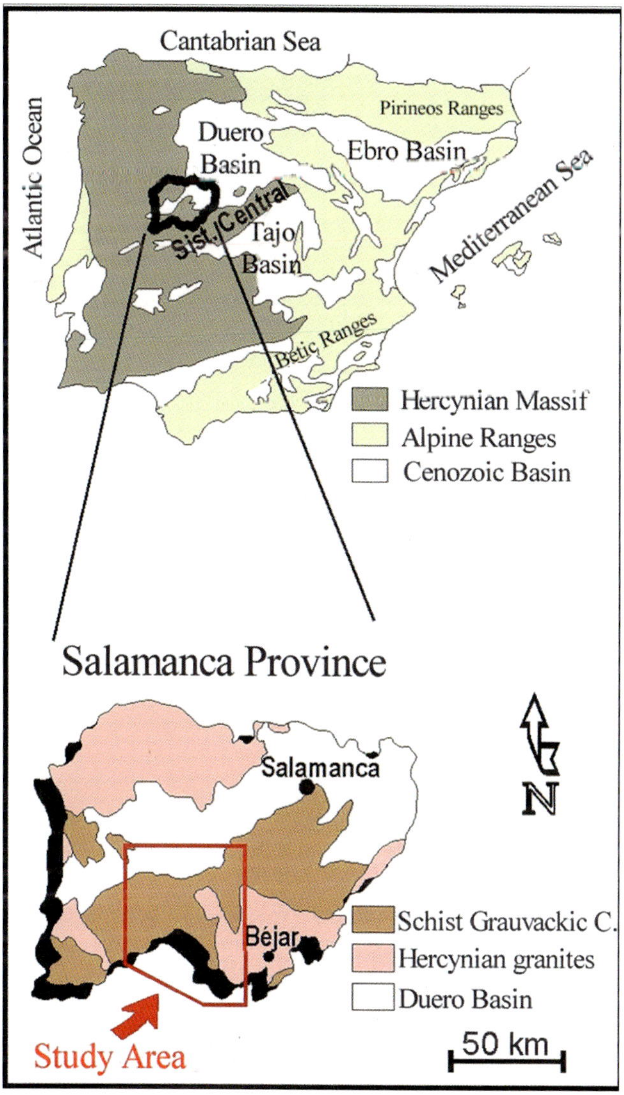

Fig. 5.1 Study area

including organic content, whether natural or otherwise. B is the material in suspension that can be separated by filtration. C is the oxygen dissolved in water. D is the electrical conductivity at 18°C; it measures the concentration of inorganic salts. The range of values of this index varies from 0 for very poor value to 100 for optimum values.

2. Map of Aquifer Vulnerability to contamination (Fig. 5.3c), drawn up from the mapping of hydrogeological units, taking into account their lithological and

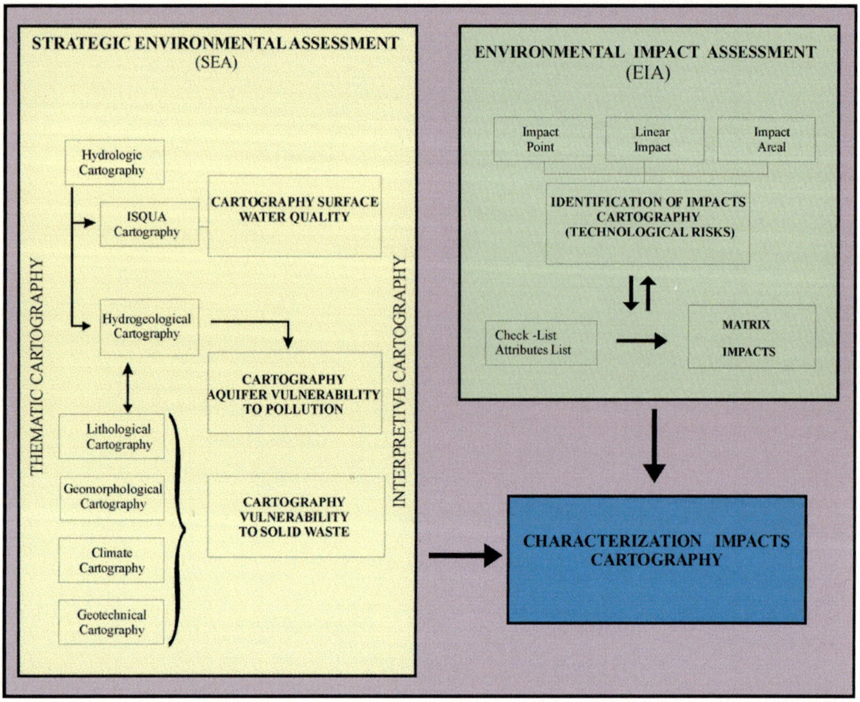

Fig. 5.2 Methodological diagram

hydraulic parameters. The GOD method was used for this purpose (Foster et al. 2003), taking into account three main factors: hydraulic containment levels (G), occurrence of overlying substrate (O) and distance to groundwater level (unconfined) or ceiling of the aquifer (confined) (D).

3. Map of Vulnerability to Solid Urban Waste – SUW – (Fig. 5.3d), obtained through GIS techniques (ArcGIS v.9) from the thematic mapping of the parameters that influence said vulnerability: lithology, surface hydrology and hydrogeology, geomorphology, climate and geotechnics. The lithological parameter takes into account the distribution of the lithology of the different sectors of the PNS (Martínez-Graña et al. 2004). Different features to be considered for the location of a landfill are taken into account: bedrock, structure (fractures and joints), surface formations (texture, compaction, etc.). The surface hydrology and hydrogeology parameter analyses these features in line with the type of permeability (K), based on the porosity (intergranular and fissure), solubility and fracturing (Martínez-Graña et al. 2004; Sanz et al. 2005). The limit values for waste disposal sites are included: non-hazardous waste landfills: K less than or equal to 1×10^{-9} m/s, and landfills for inert waste: K less than or equal to 1×10^{-7} m/s. (Williams 1998).

The geomorphological parameter helps to understand the morphological features that affect the substrate, the surface formations and the slope gradient to

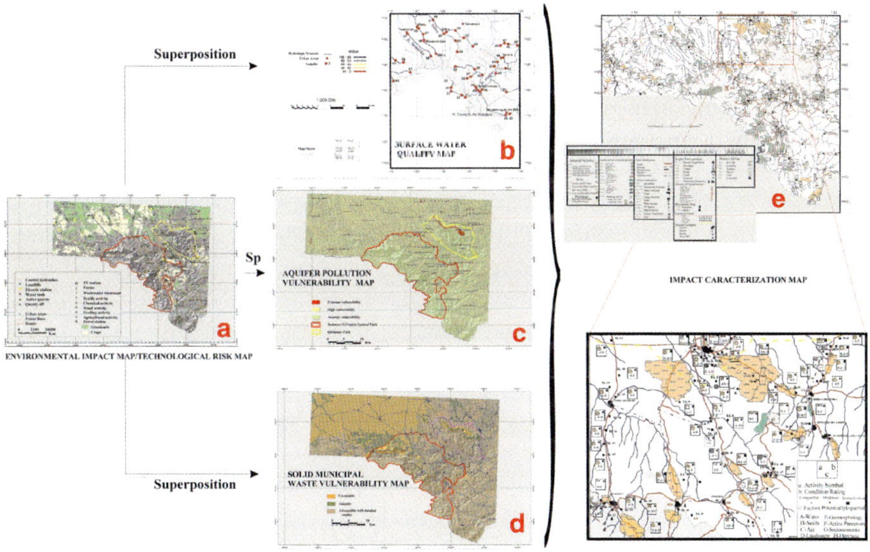

Fig. 5.3 Impact identification map (**a**) which is superimposed on the surface water quality map (**b**) aquifer pollution vulnerability map (**c**) solid municipal waste vulnerability map (**d**) obtaining the impact characterisation map (**e**) cartography details and interpretive diagram

be considered for its location (Martínez-Graña et al. 2006). This map will help to identify the active processes (risks) that determine the hazardous areas for the location of the landfill or waste treatment facility. The climate parameter analyses the rainfall distribution, as it favours the formation of leachate likely to contaminate water and soils. The Stormwater Aggressiveness map was used (Martínez-Graña et al. 2004) which was carried out based on the modified Fournier index (Fm). This mapping helps to gauge the effect of rain on the landfill and its impact on restoration (revegetation of slopes). Finally, the geotechnical parameter allows accurate planning in the operational and restoration phase of a landfill by analysing existing loan materials in the area to be used in these tasks. To do so, it is necessary to know the rippability of materials for the possibility of covering or whether it is necessary to excavate a basin to achieve a longer useful life. This variable was obtained on the basis of lithological, geomorphological and hydrological features.

5.3 Results and Conclusions

From analysing the interpretive mapping, it can be seen that the surface water quality map shows that some rivers (Alagón, etc.) present a lower water quality than their tributaries. This is due to localities dumping their urban waste water (Sotoserrano, Valero, San Miguel de Valero) as well as industrial waste (food industry, agriculture) into the channel, reducing the water quality by discharging unprocessed effluent. The lower waters corresponding to the river Yeltes present a slight decrease in quality caused by mining activities located on the plains and river terraces, creating an increase in turbidity (suspended solids) and incidence of waste typical of extraction activity (processes and machinery).

The Aquifer Vulnerability to Pollution map shows an extreme vulnerability in carbonate outcrops, where the dissolution processes of limestone and dolomite can delimit the land most vulnerable to contamination. These aquifers are extremely vulnerable to the direct and rapid spread of contaminants in fissures and cavities in carbonate rocks. In territorial land use planning, the introduction of polluting activities such as solid or liquid (urban and/or industrial) waste disposal, either on the surface or buried, should not be permitted in these sectors. Inappropriate farming practices and activities must be monitored, and the land disturbed by active or abandoned extraction activities (mining) restored. The areas with high vulnerability correspond to quaternary material, whose aquifers are highly vulnerable to the entry of pollutants from rivers, streams or direct infiltration. This pollution mainly affects free alluvial aquifers. Their ability to self-purify organic and bacteriological contamination is limited. As in the previous case, the installation of human activities likely to affect the quality of aquifers containing these quaternary formations should be closely monitored. Finally, the vulnerability is average mainly in areas where the land alternations are permeable and less permeable (detrital units) with a high capacity for bacteriological purification but low for chemical contaminants. In the second place, it may occur in impermeable igneous and metamorphic lithologies, being limited to surface water pollution and sectors of cracks and alterations.

The Vulnerability to Solid Urban Waste map shows the sectors suitable and unsuitable for landfill construction a priority, and sectors that may be favourable with detailed studies (DS), such as granite areas, to determine the saprolite thickness and degree of alteration. According to these parameters, the areas of greatest territorial acceptance of this type of human activity (landfill) are currently located near urban areas.

In second place, the Impact Identification map is superimposed on the Surface Water Quality, Aquifer Vulnerability to Pollution and Solid Waste Vulnerability maps, noting the effects and problems of different human activities in each area and their potential impact on different factors (water, soil, air, landscape, geomorphology, active processes, socioeconomic elements and heritage). The Impact Characterisation map is obtained this way, showing the absorption capacity of the different locations of human activities, the variety of types of effects (industrial, agricultural, livestock, landfills, etc.), the degree of impact (compatible, moderate, severe-critical) and the factors susceptible. The negative effects of livestock activities on soil and water resources observed are severe, compounded by a moderate degree of impact on the landscape due to the presence of electrical and communication infrastructures.

This methodology is of great interest as it provides reliable results quickly and economically, suitable for application both in rural areas with poor resources and the management of protected natural areas, constituting a useful tool to understand the state of the natural environment and in particular to analyse the impact on water quality and identify pollution sources, paving the way for further studies to identify specific problems and their solutions.

Acknowledgments Projects CGL 2008-03998/BTE, SA041A08 (JCyL) and CGL2008-04000/BTE.

References

Cubillo F (1986) Situación actual dc la calidad de las aguas en los rios de la Comunidad de Madrid (Enero 1986). Comunidad de Madrid. Consejería de Obras Públicas y Transportes. Dirección General de Recursos Hidráulicos

de Bustamante I (1989) Aspectos metodológicos en estudios de calidad del agua. Henares. Rev Geol 3:25–36

Foster S, Hirata R, Gomes D et al. (2003) Protección de la Calidad del Agua Subterránea: guía para empresas de agua, autoridades municipales y agencias ambientales. Banco Mundial, Washington

Martínez-Graña A, Goy JL, Zazo C et al. (2004) Caracterización Geotécnica de los Espacios Naturales de Las Batuecas-S de Francia y Quilamas. Cartografía básica para la ordenación del territorio. Geogaceta 36:67–70

Martínez-Graña A, Goy JL, Zazo C et al. (2006) Environmental cartography aplicable to waste management of the Natural Protected Environment « Las Batuecas-Sierra de Francia » (Salamanca- Spain). Geogaceta 40:311–314

Sanz JM, De Bustamante I, Goy JL, Mateos J (2005) Cartografía Hidrológica del Paisaje Protegido de El Rebollar (Salamanca). En: Agua, Minería y Medio Ambiente. Libro Homenaje al prof. D. Rafael Fernández Rubio. IGME

Williams P (1998) Waste treatment and disposal. Wiley, England

Chapter 6
Human Effect Over the Chemical Denudation Development in the Coastal Limestone Aquifer, Havana Southern Plane, Cuba

J.R. Fagundo Castillo, J. Pajón Morejón, P. González Hernández, M. Suárez Muñoz, and C. Melián Rodríguez

Abstract Measuring of chemical denudation in karstic terrains of Cuba, dissolved $CaCO_3$ content, and water chemical estimation analysis were developed. The results demonstrate that karstic chemical denudation at Havana southern karstic coastal plain is around of 15 $m^3/km^2/y^{-1}$. The intensity of the chemical denudation improved in Havana southern coastal aquifers as a result of ionic strength and mixing water effects, the additional carbon dioxide (CO_2) generated in the sulphate anaerobic reduction process and the waste biodegradation due human activity. Several field evidences demonstrate the increases of the karstic development.

Keywords Coastal limestone aquifer • Karst terrains • Geochemistry • Cuba

6.1 Introduction

Several quantity studies in the field of geomorphology, hydrology and karst science were carried out from the past century to determine the intensity of limestone dissolution, with the aim to estimate the karst landscape transformation and to acknowledge the nature of the process.

The French geomorphologist, Jean Corbel (1959), made a great impact on the scientific community when he published the results from analysis of thousands of

J.R. Fagundo Castillo (✉) • P. González Hernández
Chemistry Faculty, Havana University. Zapata between G and Carlitos Aguirre. Vedado, PC 10400, Havana
e-mail: juan_fagundo@fq.uh.cu

J. Pajón Morejón
Museo Nacional de Historia Natural, Obispo 61, Plaza de Armas, La Habana Vieja, La Habana, Cuba, CP: 10 100

M. Suárez Muñoz • C. Melián Rodríguez
Institute for Applied Sciences and Technology, Havana, Cuba

H. Farfán González et al. (eds.), *Management of Water Resources in Protected Areas*, Environmental Earth Sciences, DOI 10.1007/978-3-642-16330-2_6,

field samples. He concluded that temperature performs the main control of limestone solution due to the inverse solubility of the CO_2. He found that the chemical denudations were more intense in the cold than the hot regions for a given annual rainfall. These results were impugned with some scientists on the basis of the geomorphological evidence in hot humid countries.

After Corbel deceased, the investigations about the karst denudation were continued by Gams (1972) and Pulina (1974). Smith and Atkinson (1976a) and other scientist found that the karst denudation rate depends on the precipitation, runoff and also on the temperature, but assumed that climatic effect is much less marked than Corbel claimed.

Between 1984 and 1989, an investigation project about the genesis and evolution of karst (PIGEK), affiliated by UNESCO, was elaborated by the Commission for Physicochemistry and Hydrogeology of Karst of the International Union of Speleology, and was selected as experimental polygons, the Silesian paleokarst (Poland), Horndsund polar region (Spitsbergen) and Pan de Guajaibón (Cuba), representative of the temperate, polar and tropical climates. The results of this research were published by Pulina and Fagundo (1992).

An indirect form to calculate the rate of the limestone solution is by means of the determination of the water hardness ($Ca^{2+} + Mg^{2+}$ as mgl^{-1} of $CaCO_3$). The water hardness depends on the CO_2 content in the soil drained by the precipitation during the infiltration in the aquifer recharge zone. Smith and Atkinson (1976b) reported mean values of pCO_2 between 0/2 % and 11.0 % in the tropical karst (with an extreme value of 17.5 %), while in the temperate climate zone they report values between 0.1 % and 3.5 % (with occasional greatest of 10.0 %), and between 0.1 % and 0/2 % in the polar climatic zone. Miotke (1974) found in the alpine tundra CO_2 contents of the 0.01–0.5 % order. Fagundo (1996) found $CaCO_3$ content between 32 and 45 mgl^{-1} at Hordsund permafrost (Spitsbergen) between 174 and 230 mgl^{-1}.

The studies about the chemical denudation development in coastal limestone regions are limited. Douglas (1968), calculates, 15.0 $m^3/km^{-2}/y^{-1}$, at Kissimmee River (Florida), and Hanshaw and Back (1980) 46/0 $m^3/km^{-2}/y^{-1}$, at Xel Ha, northern Yucatan Peninsula (Florida). González et al. (2003) reports 15.0 $m^3/km^{-2}/y^{-1}$ at Havana southern limestone coastal plain.

The aim of this paper is to show the result about the rate of limestone solution and water quality in coastal aquifers of Güira-Quivicán basin (Havana Southern Plane). This region has been submitted to a high overexploitation for agricultural irrigation and population water supply, originating in invasion of the seawater to the aquifers (Fig. 6.1).

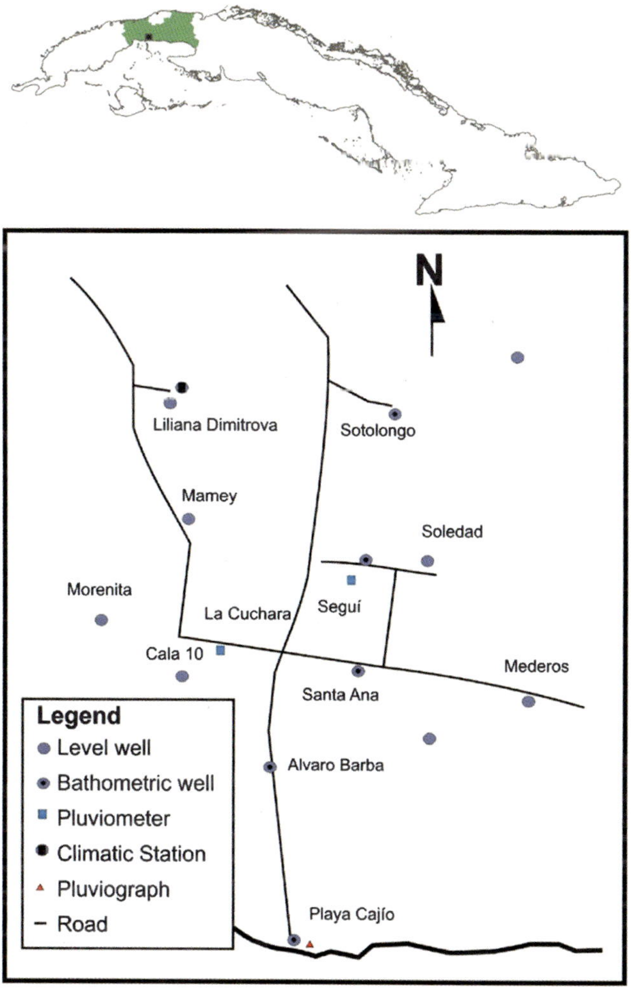

Fig. 6.1 Map showing the location of study area and the sample points

6.2 Chemical Denudation Estimate from the Study Area

6.2.1 Chemical Denudation at the Recharge Zone

The chemical denudation was calculated by the same method used by Hanshaw and Back (1980). With the aim to calculate the magnitude of the chemical denudation at the recharge zone of the Güira-Quivicán basin (Havana Southern Plain), the Rancherita well (20 m) was selected. For this purpose it was taken into account that in the study area there are no surface streams and aquifer discharge is near the line cost. The recharge of all aquifer systems can be estimated from the annual

Table 6.1 Chemical denudation in Güira-Quivicán basin (Havana Southern Plain)

Well (deep)	Calcite (mmol)	Chemical denudation (D)	
		$m^3/km^{-2}/y^{-1}$	$t\ km^{-2}/y^{-1}$
Rancherita (20 m)	2.23	31.13	77.83
Sotolongo (Díaz 50 m)	4.05	60.29	150.72
Playa Cajío (17 m)	0.268	18.86	47.6 MT
Playa Cajío (23 m)	4.241	28.85	74.62

precipitation and local evapotranspiration, and all the precipitation that infiltrates through the aquifer to dissolve the limestone to reach the calcite saturation.

The output data for chemical limestone denudation are as follows: basin surface, 1,200 km^2; coast line, 40 km; mean annual precipitation, 1,398 mm; mean annual temperature, 25 °C. The water extraction from the aquifer is as follows: irrigation, 3.5 m^3/s^{-1}; population water supply, 3.2 m^3/s^{-1}. According to the above data, the total water extraction is 6.7 $m^3/s^{-1} = 2.08 \times 10^8\ m^3/y^{-1}$.

From the balance between the mean annual precipitation (1,398 mm) and the evapotranspiration (1,050 mm), estimated by means of the Martonne formula, the mean annual recharge (assuming equal to runoff) in the basin: 349 mm = 0.349 m, was obtained. Then the mean annual total recharge (MATR) is calculated by the product of the basin surface ($1.202 \times 10^9\ m^2$) with the mean annual recharge (0.349 m), equal to $4.195 \times 10^8 m^3/y^{-1}$.

The amount of dissolved calcite in the aquifer system can be calculated assuming that the mean calcite content is equivalent to the annual dissolved calcite. How each litre of underground water which flows through the aquifer system contains the mean quantity of calcite (Table 6.1), and that the chemical denudation (D) is equal to MATR, the chemical denudation originated at the study area for each aquifer system can be calculated, multiplying the quantity of dissolved calcite by MATR:

D = (MATR)*(annual dissolved calcite)

D = ($4.195 \times 10^8\ m^3/y^{-1}$)(annual dissolved calcite. 10^3 mmol of calcite $\times$ m^{-3}) = 4.195×10^8 mol of calcite y^{-1} (annual dissolved calcite).

Dividing between the surface of the basin (1,202 km^2), changing mmol to mol, and considering that 1 mol of calcite is 100 mg l^{-1} (0.1 kg l^{-1}), the chemical denudation can be obtained for each km of coastline:

D = $3.49 \times 10^5 \times$ (annual dissolved calcite) mol of calcite per year for each km^2.

How 1 mol of calcite = 100 mg 1^{-1} = 0.1 kg.1^{-1}, assuming all carbonate rock to have a density = 2,5 Kgm^{-3}, and 1 metric ton (t) = 1,000 kg, then the chemical denudation per km^2 can be expressed in the term mt or m^3 as:

D = 34.9 $\times$ (annual dissolved calcite) t km^{-2}/y^{-1}

D = 13.96 $m^3/km^{-2}/y^{-1}$.

For Rancherita well,

D = ($4.195 \times 10^8\ m^3/y^{-1}$)($2.23 \times 10^3$ mmol of calcite/m^3)/1,202 km^2 = 7.783×10^8 mol of calcite y^{-1}/km^{-2}

D = 77.83 t km^{-2}/y^{-1} = 31.13 $m^3/km^{-2}/y^{-1}$.

This value is more than twice that calculated by Back and Hanshaw in Yucatán Península (37.5 t), which can be due to the difference in the mean annual rain among the compared areas (1.3 large in Güira-Quivicán as in Yucatán).

6.2.2 *Recharge of the Aquifer in the Discharge Zone*

To calculate the recharge of the aquifer at the discharge zone it is necessary to perform a balance between the water that enters the aquifer (MATR) and that coming out at the recharge zone by human activity (EXPL). Subtracting the water extracted in Güira-Quivicán basin for population supply, which is about 6.7 $m^3/s = 2.08 \times 10^8\ m^3$, from the value of the mean annual total recharge (MATR), the magnitude of mean annual discharge (MAD) can be obtained.

$MAD = MATR - EXPL = (4.195 \times 10^8\ m^3/y^{-1}) - (2.08 \times 10^8\ m^3/y^{-1})$

$MAD = 2.115 \times 10^8\ m^3/y^{-1}$

6.2.3 *Chemical Denudation at the Discharge Zone*

The underground waters through its aquifer flow path reach the discharge zone with a quantity of calcite depending on the carbonate chemical equilibrium. Elsewhere, in the mixing zone of the aquifer one additional amount of calcite can be dissolved or precipitated due to the combined effects of the mixing water and ionic strength processes, as well as the CO_2 obtained from anaerobic reduction of marine sulphate. As shown in Table 6.1, from the discharge zone of Güira-Quivicán basin, those wells can be selected which show calcite dissolution (Sotolongo 50 m, Álvaro Barba 34 m, and Playa Cajio 17 and Playa Cajio 23 m). Álvaro Barba well was disregarded even to display the highest $\Delta(Ca^{2+})$ magnitude, because the acquisition of the Ca^{2+} in this water is fundamentally controlled by the inverse cation exchange process.

Then the chemical denudation in this zone is:

D = (MAD)(annual dissolved calcite)

The results of the chemical denudation for the selected places in Güira-Quivicán basin (Havana Southern Plain) are shown in Table 6.1.

6.3 Chemical Denudation in Other Places

Table 6.2 shows the results of chemicals reported by several authors in different limestone terrains, with the objective of comparison with that calculated in Havana Southern Plain.

From the results shown in Table 6.2, it can be established that at similar altitude (lowlands, uplands and median mountains) the less denudation (32 $m^3/km^{-2}/y^{-1}$) occurs at the polar climatic zone (Hornsund), intermediate (20–38 $m^3/km^{-2}/y^{-1}$)

Table 6.2 Chemical denudation (m^3/km^2/year) in karst regions located in different climate zones

Climate zone	Relieve	Place (country)	D (m^3/ km^{-2}/y^{-1})	References
Tropical	Coastal plain	Yucatán Península (Mexico)	15.0	1
		Kissiimmee River (Florida)	27	2
		Havana Southern basin (Cuba)	31.0	3
		Matha Brae (Jamaica)	54	2
		Río Camuy (Puerto Rico)	40	2
	Lowlands and uplands	Sierra de los Organos (Cuba)	53–136	4
		Meseta del Guaso (Cuba)	28–62	4
	Median mountains	Pan de Guajaibón (Cuba)	92	4
Mediterranean	Lowlands and uplands	Upland karst (Eslovenia)	80–136	4
	Median mountains	Vercors Pre Alps (French)	44–98	4
		Venetian Pre Alps (Italy)	23–70	4
	High mountains	Central Apennines (Italy)	153–130	4
		Julian Alps (Eslovenia)	51–67	4
		SW Caucassus (Georgia)	114–139	4
Temperate warm and transitional	Lowlands and uplands	Dobregea (Bulgaria)	23	4
		Cracow-Silesian (Poland)	24–26	4
	Median mountains	Vrakanska (Bulgaria)	38	4
		Sudetes (Poland)	20–33	4
		Pirin (Bulgaria)	47	4
		Tatra Mts. (Polonia)	49	4
Cold temperate	Lowlands and uplands	Irkutsk (Siberia)	1–6	4
	Median mountains	Khamar Daban (Siberia)	22	4
	Median mountains	Tunkinsk Alps (Siberia)	13	4
Subpolar	Lowlands and uplands	Hornsud-permafrost (Spitsbergen)	11	4
	Median mountains	Hornsud – glaceares (Spitsbergen)	32	4

Reference: (1) Hanshaw and Back (1980); (2) Boggle (1971); (3) González et al. (2003); (4) Pulina and Fagundo (1992)

and at temperate climatic zone (Poland, Bulgaria), and the highest denudation (46–136 m^3 km^{-2} y^{-1}) at tropical climatic zone (Cuba). Similar higher chemical denudation than above was found at the high mountains of the Mediterranean region (51–139 $m^3/km^{-2}/y^{-1}$), and at five areas of New Britain in Papua New Guinea, where the rainfalls lie in the range of 5,700–12,000 mm/y.

References

Boggle A (1971) Corrosion by mixing o
Corbel J (1959) Erosion en terrain ca
68:97–120
Fagundo JR (1996) Química del Ag
Hidroquímica del Karst. Univ. Gran
Gams I (1972) Effect of runoff on corro
Res Gp GB 14(2):78–83
González Hernández P, Suárez Muñoz
(2003) Denudación química en la cue
Protección Ambiental. ISCTN 4:363-
Hanshaw BB, Back W (1980) Chemic
groundwater dissolution. Geology 8:
Miotke FD (1974) Carbon dioxide and
processes. Abh Karst Hohlenkunde A
Plummer LN (1975) Mixing of seaw
142:219–236
Pulina M (1974) Denudacja Chemiczna
PAN, Warszawa, p 159
Pulina M, Fagundo JR (1992) Tropic
Geographia Polonica Warsawa 60:19
Smith DI, Atkinson TC (1976a) Proce
Derbyshire E (ed) Geomorphic and cl
Smith DI, Atkinson TC (1976b) Proce
Derbyshire E (ed) Geomorphic and cl

6.4 Human Activity and Karst Develop

The karstic aquifers are highly vulnerable to contamination. The main anthropogenic impacts in the Cuban unconfined coastal limestone aquifers are related to intense extraction of water for agricultural irrigation and population supply, the use of fertilizer and pesticide in the agricultural plantation, the deposition of urban sewage, and wastes from the sugar, citric and other industries.

The overexploitation of the aquifers induces an extensive intrusion of the seawater inland, of the order of 10 km in Havana Southern Plain (González et al. 2003). This impact increases the limestone dissolution and the hardness of the water as a result of the mentioned ionic strength process effect. The use of fertilizer and pesticides in agriculture and wastes from the metal industry create a relatively high amount of ammonium, nitrate, nitrite and organic pollutant, heavy metals, but does not contribute to karst development. However, the domestic effluents and biodegradable wastewater produces a large amount of CO_2, H_2S and CH_4 in anaerobic reducing conditions. The first of those gases can produce an intense increase of limestone dissolution.

In some sugar and run industries of the Havana and Matanzas provinces, collapses, explosions and fires have occurred, which have been attributed to production of CO_2 and CH_4. In water sampled in a well near a run industry, 92.2–111.3 mg l^{-1} CO_2 and 94.2–160.7 mg l^{-1} H_2S contents have been reported.

6.5 The CO_2 Soil Contents

In the atmosphere the pCO_2 lies between 10^{-4} and $10^{-3,48}$ tam according to the altitude and latitude, and near $10^{-3,15}$ tam in urban zones; however, the pCO_2 in the soil atmosphere is greater as a consequence of organic release in the rooting zone.

Different pCO_2 values in soils from different latitudes have been reported in the literature: Mite (1974) found $10^{-3,0}$–$10^{-1,5}$ and $10^{-4,3}$–$10^{-3,4}$ in soil of plain and mountain soils (Rocky Mountain, Canada), respectively, and $10^{-4,0}$–$10^{-3,3}$ in Alpine tundra. Smith and Atkinson (1976a) found $10^{-2,7}$–$10^{-0,8}$ pCO_2 values in tropical soils (Puerto Rico and Jamaica), and $10^{-5,3}$–$10^{-3,4}$ in Alaska tundra.

In general, the highest soil pCO_2 and $CaCO_3$ content of the water has been found in tropical rather than in temperate and the lowest in polar terrains (Table 6.3).

6.6 Final Remarks

Taking into account that in coastal aquifers the carbonate dissolution can increase with the combined effect of ionic strength process, the mixing water, and the additional CO_2, generated by both natural and anthropogenic anaerobic reduction

Table 6.3 Mean CO_2 in soils an
1996)

Type of soil (region)
Tropical (San Marcos basin, Cuba
Tropical (lowland)
Tropical (litoral karst)
Permafrost (Spitsbergen)

[a]Calculated from chemical equilib

of organic matter, attention
future karst development in
Plummer (1975) demonstrate
aggressive from 10 % to 80
order originated by the was
climatic change, the seawate
especially in overexploited ac
can be affected.

6.7 Conclusions

The chemical denudation ob
quantitative measure of the gec
obtained and others reviewe
(lowlands, uplands and medi
occurs at the polar climatic zc
and at the temperate climatic
occurs (46–136 $m^3/km^{-2}/y^{-1}$)
chemical denudation than abov
nean region (51–139 m^3/km^{-2}
Guinea, where the rainfalls lie

The highest soil pCO_2 and
tropical rather than in temperat

By means of water-rock in
open and closed systems conc
contents were obtained than th
the Havana Southern Plain bas
found in open system condit
contents are of the order gener

Chapter 8
Characterization of the Vulnerability to the Contamination in Defined Areas of the Vento Aquifer, Validated with Geophysical Methods

I. Pedroso Herrera, M.J. Fundora, F. Méndez, M. Guerra, and I. González

Abstract In this paper an assessment of the specific vulnerability of the Vento aquifer by using geologic, geomorphologic and hydrologic considerations validated by geophysical methods (seismic and the electrical resistivity) is carried out. The application of geophysical methods allowed the determination of the presence of thin layers of clays that could protect the underlying aquifer and to know the thickness of the intermediate conductive layer, taking into account that to greater conductance greater protection of the aquifer. It also allows the evaluation of the presence of fractures, cracks and faults that are more vulnerable in the aquifer. Once obtained, the intrinsic vulnerability, the specific vulnerability of sectors of the basin, was obtained by means of the combination of the intrinsic vulnerability map with the zoning map of the potential contamination by heavy metals obtained by using techniques of environmental magnetism through the measurements of magnetic properties of soils in the field and laboratory.

Keywords Aquifer vulnerability • Vulnerability map • Heavy metals contamination • Environmental magnetism • Vento aquifer • Cuba

---

I. Pedroso Herrera (✉)
Agencia de Medio Ambiente del CTMA, Calle 20 esq. 18A Miramar, Playa, La Habana
e-mail: ida@ama.cu

M.J. Fundora
Department of Geosciences, Technical University of Havana "Jose A. Echeverria", MES, Calle 116 S/N esq. a 127, Marianao 15, La Habana, Cuba

F. Méndez
Department of Geology and Geophysics, Faculty of Geology, University of Pinar del Rio, MES, Pinar del Rio, Cuba

M. Guerra • I. González
Institute of Geophysics and Astronomy, CITMA, 212 N0 2906 esq. 31, Rpto "La Coronela", La Lisa, Cuba

H. Farfán González et al. (eds.), *Management of Water Resources in Protected Areas*, Environmental Earth Sciences, DOI 10.1007/978-3-642-16330-2_8,

8.1 Introduction

The Vento aquifer is one of the four basins that supply water to the Cuban Capital and the one that gives the greatest volume of water (8.8 m^3/s) that it consumes. To be a non-confined karstic aquifer, the hydraulic groundwater resources exhibit a high vulnerability to contamination. At the basin there are more than fifty polluting sources, some of them contaminate the soils with toxic metals. If monitoring was performed by the conventional chemical techniques, it would be very expensive. This expense could be reduced by using measurements of the magnetic susceptibility of the soils as proxy indicators of probable contamination in those areas where intrinsic and specifically larger vulnerabilities are expected in the contamination by heavy metals. To achieve a more precise evaluation of the intrinsic vulnerability, estimated by geologic and geomorphologic methods as well as of the specific vulnerability to heavy metals, a geophysical investigation with the following objectives was carried out:

1. To specify the characteristics of the geological section unto the depth of 15, 00–20, and 00 m, as well as to estimate the main physical-mechanical characteristics for the different types of rocks and present soils in the area by using a seismic method.
2. To define the presence of faults, fractures and crackings, as well as to determine the thickness of the Unsaturated Zone (USZ) and the clay content, by using electrical methods of resistivity.
3. To evaluate the specific vulnerability of the aquifer starting from the combination of the cartography of the contamination for heavy metals, obtained with environmental magnetism techniques and the intrinsic vulnerability of the aquifer, by using a project SIG.

The first two objectives allow the estimation of important elements necessary to evaluate the intrinsic vulnerability of the aquifer to the contamination.

8.2 Materials and Methods

The cartographic materials used were the geologic, lithologic, geomorphologic and hydrogeologic maps of the basin.

The lithologic composition allowed the differentiation of the two main hydrogeologic units, a central one very karstified and one that constitutes the main aquifer of the basin and which is evaluated as a holokarst by its lithologic thickness; and another one is located in the borders of the basin, which is less karstified and is evaluated as a merokarst for its thickness and lithologic composition. To complete the required information, in order to evaluate the intrinsic vulnerability, the geophysical works of two seismic lines were carried out. The results of the electric profiles and Vertical Electrical Soundings carried out in previous works by the National Institute of Hydraulic Resources, were reinterpreted.

8.2.1 Seismic Method

Seismic lines in two areas were carried out, one of them characterized by marls rocks and the other by calcareous ones in order to determine the physics and mechanics parameters of these rocks for different degrees of conservation.

The seismic observations were carried out for the refraction variant using a base line of 24 channels, with the typical methodology followed for these works (Gurvich 1975). For the interpretation of the seismic data the software Winsim 9 was used and the following procedures: Method of interception points, interpretation of Retard (Delay – Method) and in some cases the Widespread Method of the Reciprocal ones was used (General Reciprocal Method or GMR; Mareš 1984).

8.2.2 Geoelectrical Methods

With the acquired data using the Vertical Electric Sounding (VES) and the combined electrical profiling variants for the measurements of the electric resistivity methods (Orellana 2007), the corresponding electric profiles were elaborated as well as maps of isolines for the apparent resistivity, with which resistive variations of the section could be pointed out, so much vertical as horizontally correlating this information with some anomaly. Seven profiles were carried out.

8.2.3 Environmental Magnetism

To assess the specific vulnerability of the Vento aquifer, an outline for the zoning of the pollution of the soils by heavy metals in the basin was used. This was built up by starting from surveys of the magnetic susceptibility on a 250 by 250 m grid to a scale of 1:250,000 as well as other surveys carried out on grids of different scales around some industries, roads and some others possible sources of pollution (Pedroso Herrera et al. 2005).

The measurements were processed statistically according to the standard procedure for the magnetic properties of the rocks (Mareš 1984; Pedroso Herrera et al. 2005; Pedroso et al 2004). To estimate the values of the magnetic susceptibility of the soils that characterize the non-polluted areas it was considered a value threshold by means of the processing of magnetic susceptibility measurements in non-polluted areas, taking as such those far away ones more than 5 km of the contamination sources. To characterize the susceptibility of the polluted soils, the measurements in areas close to the potential sources of contamination in a radius smaller than 500 m were processed statistically. For this processing the application Software STATISTICA was used for Windows version 5.0. To create the isolines map of the magnetic susceptibility, SURFER software, version 8.2 was used (Pedroso Herrera et al. 2005).

Using the Geographic Information System (GIS) elaborated in the frame of the project, the outline of the zonation of the contamination by heavy metals was made according to the cartography of the magnetic susceptibility.

8.3 Results and Discussion

8.3.1 Geologic Interpretation of the Seismic Information

8.3.1.1 Area of Antillana de Acero

The section in this area recognizes a layer of vegetable covering and underlying to this layer, there are clayed marls that have been referred by the seismic interpretation as impermeable and being developed until the depth of 3 m. Among the depths of 5–35 m, two levels of very defined karst are recognized which are composed by marl limestone or calcareous marl in an upper layer and by karstified limestone in a lower one. Starting from here one can interpreted that the area is impermeable only until the first 3 m deep.

8.3.1.2 Area of the Vocational School "V.I. Lenin"

In this section a first horizon is observed, which is formed by a vegetable layer composed of clayed sands with organic matrix and very porous and very plastic red clay that is developed until the 3 m deep. These results reveal a high permeability from the surface, as this area is very vulnerable.

The results show that the area of the Antillana de Acero is impermeable only until the first 3 m deep, while the area of the Lenin School has a high permeability from the surface, as this area is more vulnerable.

8.3.2 Interpretation of the Geoelectric Results

For each profile, a geologic and geomorphologic interpretation of the results given by the geophysical methods was carried out to estimate the vulnerability of the basin, considering the porosity, permeability and claying of the rocks, as well as the presence of possible faults. In this case it will show two cases as examples.

8.3.2.1 Profile I

In the geoelectric section prevail the calcareous marl, upward they have a greater content of clay, which is manifested in the vertical sections of the apparent electrical resistivity. Downward this marl becomes calcareous with cracks and fissures. The vegetable layer has a thickness of about 5–8 m. In this section an anomalous area is manifested, similarly it occurs with the combined electric profiling. This anomaly corroborates the supposed fault of NW – SE direction.

8.3.2.2 Geologic and Geomorphologic Considerations

The geoelectric section indicates the presence of two lithologic layers that underlie a covering thickness of up to 8 m. The upper layer is constituted by calcareous marl with a high content of clay in the upper part. The lower layer is composed of that same marl, but here it is more calcareous, karstified and has cracks and fissures. In the vertical section of this profile, due to the lithologic changes that occur, the behavior in depth for the different layers of the profile causes the conditions to be of the intrinsic vulnerability that vary in depth. In this case the vulnerability is mean.

The possible presence of a fault can benefit the favorable conditions for the infiltration, being the direction of the flow is that of the fault.

8.3.2.3 Profile II–III

In this section the marl limestone has a manifested crack that leaves making it more compact and more calcareous. It is also manifested markedly in the combined profiling and in the section of the resistivity isolines. More below in the section calcareous marls and marl exist. In the profile there is another area of interest that can be related with the first tract of the fault guided in the SW-NE direction.

8.3.2.4 Geologic: Geomorphologic Considerations

The cracked marl limestone presents favorable conditions for the infiltration that occurs through the cracks. The presence of calcareous marl, more compact and more calcareous, should slow velocity of the flow that traffics the cracks; however, at the base of the section composed by calcareous marls and marls there should be another local screen that defuses the flow. The presence of faults favors conditions of the infiltration. In this case the vulnerability of the profile is high.

8.3.3 Map of Zonation of the Potential Contamination for Heavy Metals

In Table 8.1 are shown the ranges of values of the magnetic susceptibility for the classifications of levels of contamination to the zones established starting from the statistical analysis of the measurements of the magnetic susceptibility.

Table 8.1 Classification of the levels of contamination for the soils in areas of the Almendares – Vento basin and their ranges of values for the magnetic susceptibility

Classification	Range of kappa in 10^{-3} SI
Not polluted	0.0–0.627
Lightly polluted	0.628–1.65
Fairly polluted	1.66–2.6
Polluted	2.67–15.0
Very polluted	>15.1

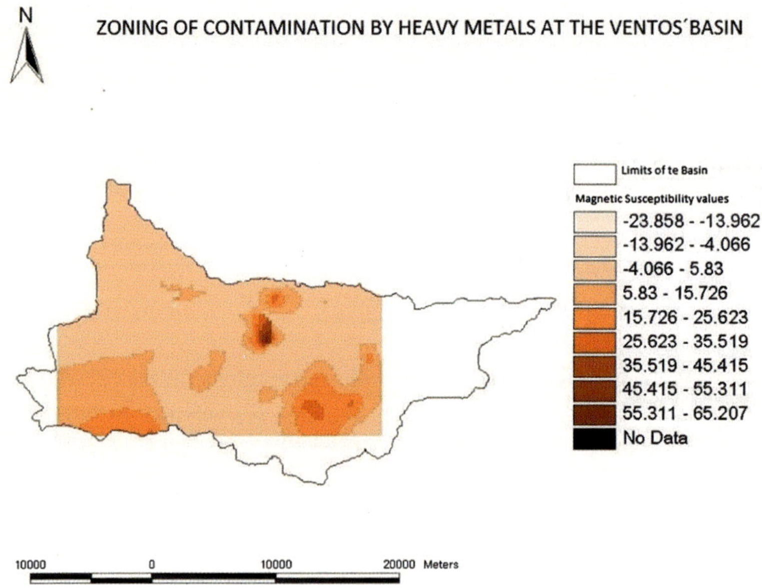

Fig. 8.1 Zoning of the pollution by heavy metals

In Fig. 8.1 is shown the map of Zoning of Potential Contamination of the soils by heavy metals in the Almendares – Vento Basin.

8.3.4 Maps of Intrinsic and Specific Vulnerabilities of the Vento's Aquifer

Starting from the geologic, geomorphologic and hydrogeologic data enriched by those obtained from the seismic method and the resistivity electrical methods, the following general characterization of the USZ could be carried out; and by using the GIS elaborated for the project, the map of the intrinsic vulnerability of the groundwater basin Vento was built. It is shown in Fig. 8.2.

The obtained map of intrinsic vulnerability was combined with the cartography of the magnetic susceptibility to evaluate the areas that had potentiality of being

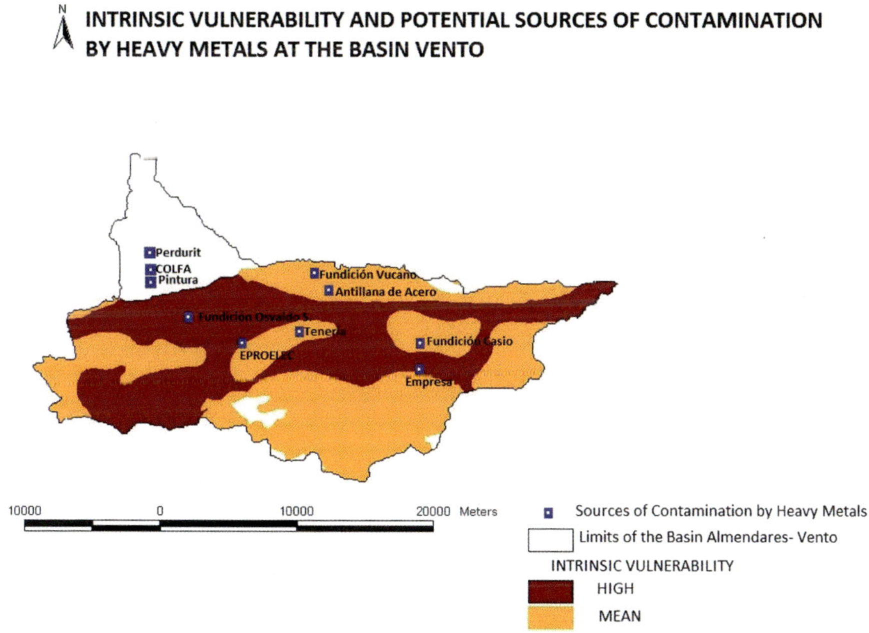

Fig. 8.2 Map of the intrinsic vulnerability for the Vento Basin

affected by contamination of heavy metals from the point of view of the vulnerability of the aquifer. Therefore, the map of specific vulnerability was obtained and it is shown in Fig. 8.3. To create this map the algebra of maps was used with the aid of a GIS.

8.4 Conclusions

The combination of several geophysical methods for the cartography of the contamination and the evaluation of the vulnerability of the aquifer revealed an understanding of the most vulnerable areas to the contamination for infiltration of the sector studied in the basin Vento.

The presence or absence of layers of low permeability in the USZ is a very important parameter when establishing strategies for the protection of the aquifers. The knowledge of the clay content that one obtains with the aid of the geophysical methods is an element that helps to evaluate this permeability.

The area of the industry "Antillana de Acero", according to the results given by the seismic methods, is impermeable only unto the first 3 m deep.

In the area of the "Vocational School Lenin", the results reveal a high permeability from the surface, as this area is more vulnerable.

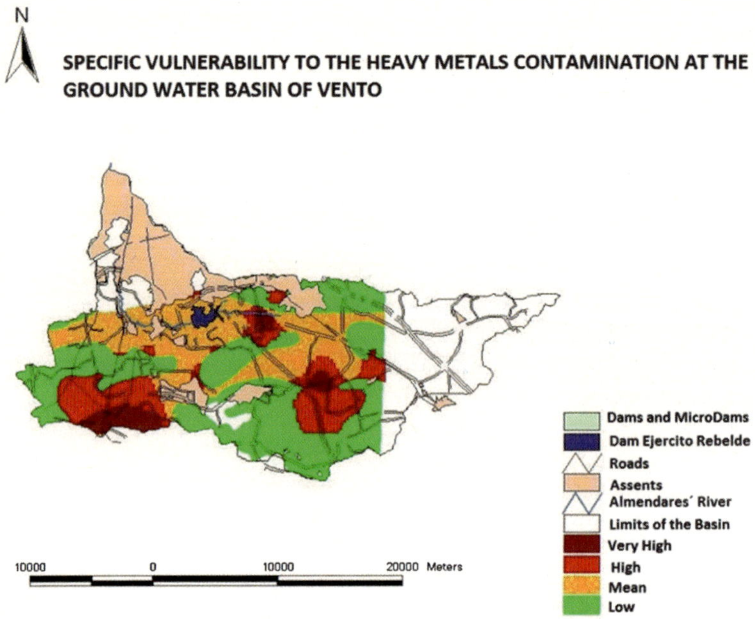

Fig. 8.3 Map of specific vulnerability for contamination by heavy metals in the basin Vento

The basin of Vento presents a high intrinsic vulnerability. For that reason when it is in the presence of potential hazards of contamination by heavy metals, it also possesses a high specific vulnerability for those polluting elements.

References

Gurvich II (1975) Sismorazviedka (Seismic Prospecting), ED. Nedra. Moscow, Russia

Mareš S et al (1984) Introduction to applied geophysics. Reidel ED, Publishing Company Holland, in co edition with SNTL Publishers of technical literature, Prague

Orellana E (2007) Prospección Geoléctrica de Corriente Continua. ED. Paraninfo-Thompson, Madrid

Pedroso Herrera I, Fundora Granda M, Jaimez Salgado E et al (2004) Diagnostico ambiental para la evaluación de áreas potenciales de contaminación por metales pesados en la cuenca del río Almendares. Final Report of Research Project, IGA, CITMA, La Habana, Cuba

Pedroso Herrera I, Fundora Granda M, Jaimez Salgado E et al (2005) Cartografía de la susceptibilidad magnética en áreas de la cuenca del río Almendares. Final Report of Research Project, IGA, CITMA, La Habana, Cuba

Chapter 9
Assessing the Vulnerability of Groundwater Pollution at Sensitive Areas by Geophysical Methods

Y. Diaz, M. Himi, J.C. Tapias, L. Rivero, X. Font, and A. Casas

Abstract The protection of groundwater has became one of the foremost environmental European policies according the Orders on the Protection of Water Against Contamination Directive 2000/60/CE of the European Parliament and the European Union Council, and more concretely with the Directive 2006/118/118/CE, related to groundwater protection against the pollution and degradation. Traditional methods for assessing vulnerability include soil surveys, drilling and analyses of logs from wells with the objective to characterize the thickness, hydraulic properties and lateral extent of the protective layers. Disadvantages of such investigations are that they can be labour-intensive and expensive. In addition, the measured parameters may have high spatial variability making accurate characterization over large areas difficult. As a result, a numerical index of protection can be assessed from the longitudinal electrical conductance parameter derived from electrical resistivity surveys (VES, ERT or any other electrical or EM method), that can be more accurate and reliable than any other vulnerability index assumed only by visual inspection or interpolated from sparse borehole data.

Keywords Groundwater protection • Vulnerability mapping • Geophysical methods • Land management • Spain

---

Y. Diaz • M. Himi • L. Rivero • X. Font • A. Casas (✉)
Water Institute, Geochemistry, Petrology and Geological Prospecting Department, Faculty of Geology, University of Barcelona, Barcelona, Spain
e-mail: albert.casas@ub.edu

J.C. Tapias
Water Institute, Soil Science Laboratory, Faculty of Pharmacy, University of Barcelona, Barcelona, Spain
e-mail: jtapias@ub.edu

H. Farfán González et al. (eds.), *Management of Water Resources in Protected Areas*, Environmental Earth Sciences, DOI 10.1007/978-3-642-16330-2_9,

9.1 Introduction

During the last years different methods for assessing the vulnerability have been developed for which the main objective is to favour an adequate land management and help to avoid the pollution of groundwater because once polluted the remediation process is time-consuming and expensive, if not impossible. This is of particularly concern for the aquifers located in protected areas, where, in general, scarce direct information is available about the subsoil conditions, and therefore is needed to complement the surface geological data, other subsoil data, as for instance indirect geophysical methods.

The near-surface environment acts as a filter and buffer for contaminants introduced from the surface by anthropogenic activities. For this reason there is a great need to improve the understanding of the shallow subsurface, taking into account the increasing demand for vulnerability maps, which illustrate the exposure of aquifers against pollution. These maps are designed to show areas of greatest potential for groundwater contamination on the basis of local subsurface conditions.

9.2 Vulnerability Assessment

Groundwater must be protected against pollution sources, such as landfills, pit latrines, industries, agrochemicals, etc.; and vulnerability maps can be used to illustrate the exposure of aquifers to pollution. Groundwater vulnerability has become a central concept in assessing the risk of groundwater pollution (Foster 1987). The term vulnerability is applied to a group of essential characteristics that determine the degree of protection that natural environment provides to an aquifer affected by a polluting load. Many methods for assessing the vulnerability of groundwater have been developed. One most widely used is DRASTIC (Aller et al. 1987), but the best adapted for integrating geophysical information is the Aquifer Vulnerability Index (AVI) method (Van Stempvoort et al. 1993) and has been applied extensively for planning purposes in many areas of the world (Anderson and Gosk 1989).

9.2.1 AVI Method

The AVI method developed in Canada by Van Stempvoort et al. (1993) uses two variables to formulate a vulnerability index: the thickness of each sedimentary layer above the uppermost saturated aquifer (h) and the estimated hydraulic conductivity of each of these layers (k). The vulnerability index is then expressed in terms of the sum of the hydraulic resistance (c) of each layer:

$$c = \sum_{i=1}^{n} \frac{h_i}{k_i} \tag{9.1}$$

Thickness of individual sedimentary layers can be taken directly from the well records. The K values for sandy material (10^{-5}–10^{-1} m/s) are several magnitudes higher than those for clayey layers (10^{-8}–10^{-6} m/s). Therefore, hydraulic resistance, as defined above, is dominated by clayey layers. Hydraulic resistance (c) has the dimension of time (e.g. years) and represents the flux–time per unit gradient for water flowing downward through the various sediment layers to the aquifer. The lower the global hydraulic resistance (c) the greater will be the vulnerability of the underlying aquifer.

This parameter has the same form that the longitudinal electrical conductance defined by Maillet (1947) as the second Dar Zarrouk parameter. Fortunately, this fact solves the ambiguity stated by the equivalence principle inherent of electrical resistivity interpretation, because it is the parameter that is independent of the chosen model.

9.3 Geophysical Approach

The purpose of electrical surveys is to determine the subsurface resistivity distribution by making measurements from the ground surface. From these measurements, the true resistivity of the subsurface can be estimated.

Unconsolidated sediments generally have even lower resistivity values than sedimentary rocks, with values ranging from about 10 to less than 1,000 Ω·m. The resistivity value is dependent on the porosity (assuming all the pores are saturated) as well as the clay content. Clayey soil normally has a lower resistivity value than sandy soil. For a better understanding of the theory of the electrical behaviour of stratified conductors, there are some fundamental parameters called Dar Zarrouk (DZ) parameters that are related to different combinations of the thickness and resistivity of each geoelectrical layer in the model (Maillet 1947).

For a sequence of n horizontal, homogeneous and isotropic layers of resistivity ρ_i and thickness h_i the DZ parameters (longitudinal conductance S and transverse resistance T) are defined respectively:

$$T = h \cdot \rho \tag{9.2}$$

$$S = \frac{h}{\rho} \tag{9.3}$$

VES sounding inversion and interpretation are inextricably limited by the ambiguity stated by the principle of equivalence that expresses that an experimental VES curve can be explained as produced by many physically equivalent models.

According to this principle, high resistivity layers can only be modelled by their T (that means the constant product is a given thickness and resistivity) whereas, low resistivity layers are better defined by the longitudinal conductance S. As an example of the sensitivity of these parameters, if a vadose zone is 20 m thick, the longitudinal conductance will vary from 2 simens in the case of a low resistivity layer (10 Ω.m) simulating a pure clay layer, to 0.04 siemens when the layer is formed by gravels or coarse sand having an electrical resistivity of 500 Ω·m.

Clay layer corresponds with low resistivities and low hydraulic conductivities, and vice versa, hence, the protective capacity of the overburden could be considered as being proportional to the ratio of thickness to resistivity – longitudinal conductance (S). The correlation between the resistivity and the hydraulic conductivity can only be considered in clean saturated sediments, whose natural fluid characteristics are considered constant (Henriet 1975).

The principle of equivalence introduces a well-known ambiguity in resistivity interpretation. A low resistivity layer (for instance, a clayey layer) sandwiched between two layers of higher resistivity will tend to concentrate current in it. The total current carried by it will be unaltered if its resistivity ρ increases; but, at the same time increase its thickness, h (within some limits), so that the ratio h/ρ is constant. In either case, a unique determination of h and ρ would be difficult, if not impossible.

Another great limitation of the resistivity sounding method is that it does not take into account lateral changes in the layer resistivity. This limitation is solved by the use of the electrical resistivity tomography method.

The relationship between soil parameters like clay content, cation exchange capacity, and vertical hydraulic conductivity to electrical resistivity enables a vulnerability assessment based on geoelectrical or electromagnetic measurements. Results of these measurements can be used as an estimate of clay content and vertical hydraulic conductivity of the non-saturated zone (Kirsch 2006). Previous applications of electrical resistivity in assessing and mapping the vulnerability of aquifers shallow to pollutants have been reported by Kalinski et al. (1993) using VES, Kirsch et al. (2003) using airborne EM method and Casas et al. (2008) using ERT, showing the potentiality of this methodology.

9.4 Results and Discussion

The methodology has been tested at two different sites of Catalonia (NE Spain) that have been classified as vulnerable zones. In the first one located at the Valls basin VES has been used because of its wide extension, whereas electrical resistivity tomography has been applied at the Empordà basin with the aim to delineate with high resolution the lateral variability of vulnerability in a protected area. Figure 9.1 shows the location of both areas placed over the zones declared as vulnerable by the Water Authority of Catalonia.

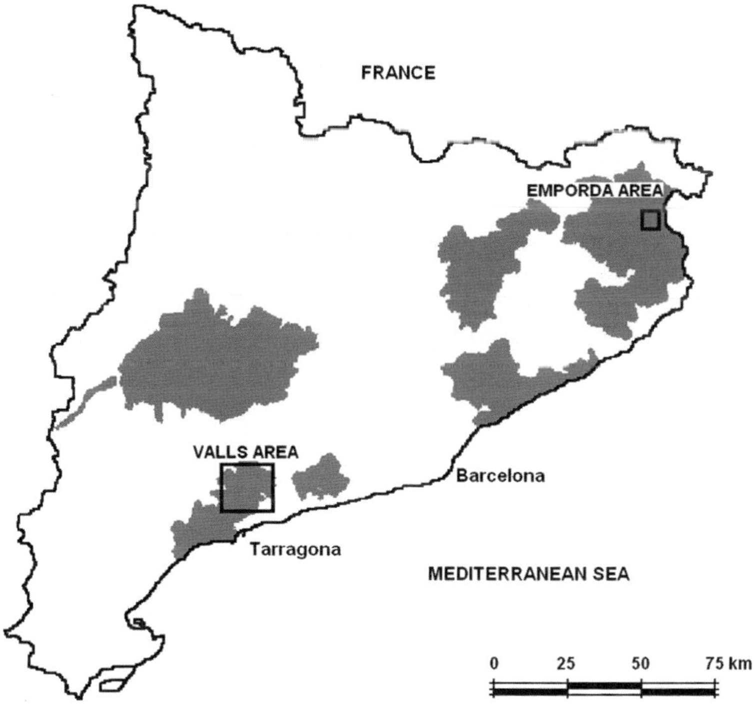

Fig. 9.1 Location of the studied areas imposed over the zones declared as vulnerable in Catalonia

9.4.1 VES Survey at Valls Basin

The study zone is located on the Alt Camp region (Tarragona), between the Gaià and Francolí rivers and about 100 km south of Barcelona city covering a surface around 320 km^2. Morphologically, there is a wide plain disposed along a northeast-southwest axis and limited by the ranges of the Catalan System.

An upper alluvial aquifer with high heterogeneity and anisotropy covers most of the basin mainly in the flat areas and the sectors close to the Francolí and Gaià Rivers where some fluvial terrace can be recognised. The alluvial deposits are composed of gravels of variable sizes with high transmissivity values. The Plio-quaternary aquifer is locally unconfined but in general the existence of clayey layers generates hydrogeological barriers.

Because more than 60 % of the surface is dedicated to agricultural labour, high amounts of nitrogenous compounds are found in groundwater from both aquifers. Historically, the mean concentration of nitrate in waters of wells from the alluvial aquifer of the Francolí River during the period 1969–1989 was around 15 mg/L but now this value has been increased to near 50 mg/L. The nitrate concentrations in the sampled wells show values that range from 3.13 y 138.3 mg/L, with the highest values located near the villages of Vila-rodona and Vilabella.

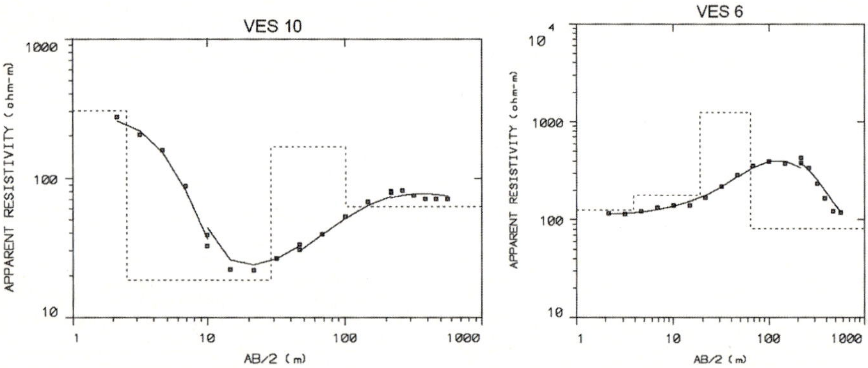

Fig. 9.2 Typical VES curve from a site well (*left*). Typical VES curve from a site where the aquifer is non-protected against pollutants (*right*)

Several resistivity surveys using VES have been conducted in the area by IGME (Spanish Geological Survey) during 1981 and 1985. The VES were executed using the Schlumberger array and the maximum current electrode spacing (AB/2) up to 1,000 m, aiming to obtain information about the thicknesses and resistivities of the shallow subsurface and also the whole aquifer section. Typical VES apparent field curves have been obtained but for brevity, only two field curves are presented here. These are soundings 6 and 10 that represent two opposite models; whereas VES 10 is a typical H curve showing a thick low-resistivity layer (Fig. 9.2, left), VES 6 show intermediate resistivity values (approximately 150 Ω·m) over the high resistivity layer (about 1,000 Ω·m) representing the unprotected aquifer (Fig. 9.2, right).

The results indicate that the subsurface consists of different geologic units such as gravel and sand, interbedded with clay layers. The main results are related to the characterization of the clay formations, which is revealed by low-resistivity values (<50 Ω m) and sand layers associated with high-resistivity values (>600 Ω m) depicted in the borders of the study zone.

According to the equivalence principle VES the inversion of the apparent resistivity values to obtain a layered model has no-unique solution, in particular for H and K type of curves. In the case of a typical H curve, like VES 10 the ambiguity is mainly when the thickness and resistivity of the low resistivity layer are analyzed, which is the most important for assessing the vulnerability in this site. Effectively, the resistivity of the low resistivity layer that ranges from 18 to 26 Ω m, and also their thickness that ranges from 14 to 18 m, keeps the error fit below 5 % (Fig. 9.3, left). Nevertheless, in any case the longitudinal electrical conductance is about 0.7 siemens, and then S values have been used for any VES.

The map of longitudinal conductance (Fig. 9.3 right) illustrates the protective capacity of the overburden layer on the confined aquifer. In this illustration, values of $S > 1.0$ siemens would indicate zones in which the confined aquifer would be protected. In comparison, values of $S < 1.0$ siemens would indicate zones of probable risks of contamination.

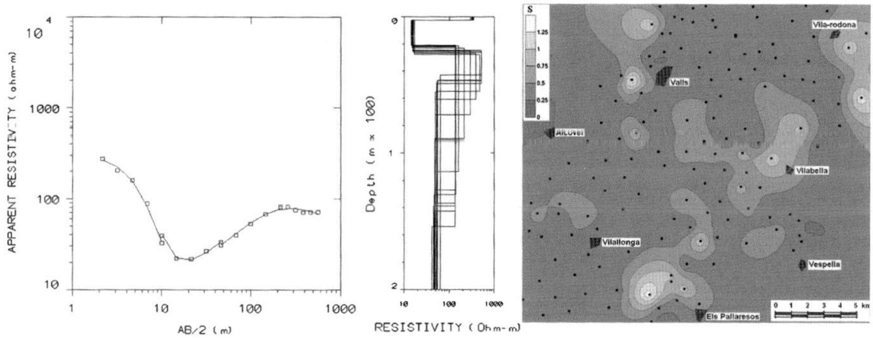

Fig. 9.3 Equivalent models for a typical H type VES curve (*left*). Longitudinal conductance map of the Valls basin (*right*)

9.4.2 ERT at Emporda Basin

The study area, located 100 km north of Barcelona, is constituted by a wide alluvial plain, where two overlaying aquifer units are recognized. The upper aquifer, which is less protected against pollution, is considered as a wide, unconfined aquifer partially covered by silt and clay deposits that act as a protective layer. Their lithological and hydrogeological characteristics fluctuate laterally, although fine-grained sands and gravels prevail.

The thickness of the upper aquifer ranges from 12 to 22 m from inland to the coastline. The estimated permeability is 500 m/day near the Muga River plain, and about 200 m/day in the rest of the aquifer. The piezometric level is generally very shallow with some surfaces covered by wetlands, making up a protected natural park named 'Aiguamolls de l'Empordà'.

To assess thickness and lateral extent of the protective layers over the aquifer, a geoelectrical survey was conducted in the vicinity of the Sant Pere Pescador, where several 2-D electrical resistivity tomography (ERT) profiles have been taken. The program RES2DINV (Loke 2002) was used to invert the apparent resistivity values to a resistivity model section. In this method, the subsurface is divided into cells of fixed dimensions for which the resistivities are adjusted iteratively until an acceptable agreement between the input data and the model responses is achieved, based on a non-lineal optimization technique by least-squares fitting (Loke and Barker 1996).

The result of a standard smooth inversion of any ERT profile shows recognizable layers of low resistivities at shallow depths. In each 2-D geoelectrical section, clayey dominated layers can be easily identified due to the electrical resistivity contrast with permeable sandy layers. Resistivities range between 10 Ω.m for the clayey layers and 60 Ω.m for the unsaturated sandy layers. Three of the most representative profiles are discussed here. The inverse model of profile 1 (Fig. 9.4 top) shows an almost horizontal interface between low resistivity values in the upper part and highest resistivity values at the bottom of the section. The thickness of the layers

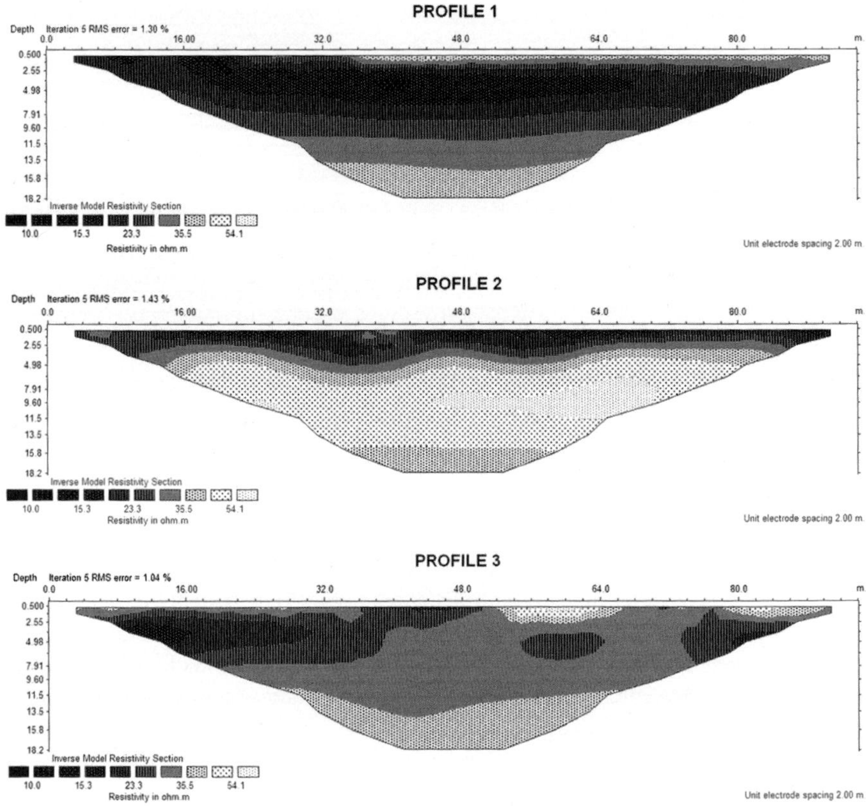

Fig. 9.4 Inverted resistivity cross-sections obtained at the Alt Empordà basin showing different vulnerabilities as a result of the thickness and lateral continuity of a low resistivity (clay dominated) layer depicted in *blue*

having resistivities lower than 20 Ω.m is more than 10 m. Conversely, the cross-section displayed in the inverse model of profile 2 (Fig. 9.4 middle) shows a thin and discontinuous low resistivity cover having a maximum thickness of 3 m. The discontinuous character of the low resistivity layer (interpreted as a clayey layer) is particularly evident on profile 3 (Fig. 9.4 bottom). In this case, the geoelectrical cross-section shows how the conductive layer, perfectly visible at the left side on the profile, decreases and almost disappears about 30 m from its origin.

9.5 Conclusions

The vadose zone over shallow aquifers plays a crucial role in protecting shallow aquifers against pollutants. However, a detailed description of soil spatial distribution is generally difficult to achieve, as soil investigations preformed by drilling auger holes or trenching and laboratory analysis, are faced with both methodological and financial constraints.

Vertical electrical soundings at regional scale, and particularly the electrical imaging technique, are very well adapted to provide precise information about depth, thickness and lateral continuity of clay-dominated layers that act as natural barriers. Besides, geophysical methods are fast, cost-effective and non-destructive, that means they provide subsoil information without drilling that could represent the development of artificial pathways for transmission of pollutants to the aquifers.

Acknowledgments This work has been partially supported by the Spanish Ministry for Science and Innovation through the research project "Applying electrical resistivity tomography for assessing groundwater vulnerability of aquifers in sensitive areas" (307165/2003–5) and the project "Waste water treatment and reuse for a sustainable management, TRAGUA" (CSD2006-00044) of the Consolider Program.

References

Aller L, Bennett T, Lehr JH, Petty RJ, Hackett G (1987) DRASTIC: a standardized system for evaluating ground water pollution potential using hydrogeologic settings. NWWA/EPA Series. EPA-600/2-87-035. Ada, Oklahoma

Anderson LJ, Gosk E (1989) Applicability of vulnerability maps. Environ Geol Water Sci 13:39–43

Casas A, Himi M, Díaz Y, Pinto V, Font X, Tapias JC (2008) Assessing aquifer vulnerability to pollutants by electrical resistivity tomography (ERT) at a nitrate vulnerable. Environ Geol 54:515–520

Foster SSD (1987) Fundamental concepts in aquifer vulnerability, pollution risk and protection strategy: In: van Duijvenbooden W, van Waegeningh HG (eds) In: Vulnerability of soil and groundwater to pollutants, Proc Inf 38, TNO Committee on Hydrological Research, The Hague, pp 69–86

Henriet JP (1975) Direct applications of the Dar Zarrouk parameters in ground water surveys. Geophys Prospect 24:344–353

Kalinski RJ, Kelly WE, Bogardi I, Pesti G (1993) Electrical resistivity measurements to estimate travel times through unsaturated ground water protective layers. J Appl Geophys 30:161–173

Kirsch R (2006) Groundwater protection: vulnerability of aquifers. In: Kirsch R (ed) Groundwater geophysics. Springer, Berlin/Heidelberg, pp 459–470

Kirsch R, Sengpiel KP, Voss W (2003) The use of electrical conductivity mapping in the definition of an aquifer vulnerability index. Near Surf Geophys 1(1):13–19

Loke MH (2002) RES2DINV ver. 3.50. Rapid 2-D resistivity and IP inversion using the least square method. Geotomo Software, Penang

Loke MH, Barker RD (1996) Rapid least-squares inversion of apparent resistivity pseudosections by a quasi-Newton method. Geophys Prospect 44:131–152

Maillet R (1947) The fundamental equations of electric prospecting. Geophysics 12:529–556

Van Stempvoort D, Ewert L, Wassenaar L (1993) Aquifer vulnerability index. A GIS-compatible method for groundwater vulnerability mapping. Can Water Resour J 18(1):25–37

Chapter 10
Mapping Groundwater Vulnerability in Guanahacabibes National Park, Western of Cuba

C. Díaz-Guanche, C. Aldana-Vilas, and H. Farfán-González

Abstract Guanahacabibes National Park is in the western part of Cuba and is characterized by a very karstified relief in a coastal plain. Conformed by Pliocene-Quaternary rocks, the karst is represented principally by a development of dolines, cenotes, extensive karrenfield, and notable cavernament. The natural and ethno-ecological values of this area are the principal criteria for defining the diverse management categories that are confluent in these territories. In this area, studies on seawater intrusion have been made previously, but the aquifer vulnerability to the human activities is not realized yet. In this paper, a map on the vulnerability to pollution was made as a tool to support the management plan. Due to the scarce database, the EPIK method has been applied. The results show that high vulnerability prevails, and low vulnerability is absent.

Keywords Aquifer vulnerability mapping • EPIK method • Cuba

10.1 Introduction

A groundwater vulnerability map is an efficient tool for management of water and land use in any territory. Their principal utilities are related to the definition of protection zoning and land-use planning, based on the relationship between water protection and economic interests. However, the special characteristics of karst aquifers, such as rapid infiltration and transport over long distances make them

C. Díaz-Guanche (✉)
Depto. Geology, University of Pinar del Río, Calle Marti 270 Final, 20100 Pinar del Río, Cuba
e-mail: carlosdg@geo.upr.edu.du

C. Aldana-Vilas
National Center of Protected Areas, Ciudad de la Habana, Cuba, Havana

H. Farfán-González
Viñales National Park, Vinales, Cuba

H. Farfán González et al. (eds.), *Management of Water Resources in Protected Areas*, Environmental Earth Sciences, DOI 10.1007/978-3-642-16330-2_10,

more vulnerable to contamination than other hydrogeological environments (Ravbar and Goldscheider 2009; Zwahlen 2004). The assessment of the intrinsic vulnerability of karst areas is difficult as the type and stage of karst development and the related underground discharge behavior are difficult to determine and quantify (Cucchi et al. 2008).

On other hand, accessibility to the database is very important criteria to determine the method to be applied. For this reason, and for the necessity of defining the vulnerability of anthropogenic activity in the areas of Guanahacabibes National Park, the EPIK method is selected. Although this method shows limitations (Goldscheider 2002; Ravbar 2007), it has been applied under several contour conditions at world level (Doerfliger et al. 1999; Tripet et al. 2000; Goldscheider 2002; Petrič and Šebela 2004; Barrocu et al. 2006; Farfán et al. 2009).

10.2 Geological and Geomorphological Setting

The Peninsula of Guanahacabibes constitutes an example of the recent geologic and geomorphologic development with wide presence of Pliocene and Quaternary carbonated rocks that have been modeled mainly by karst processes.

Stratigraphically, the base of the cut is mainly terrigenous, with some presence of carbonated rocks (San Cayetano Formation and Esperanza Formation) of terrigenous-carbonated composition with isolated intrusive bodies. Neither of these two formations, belonging to the structure-facial Area of Guaniguanico, appears in the area. Later there was a disagreement that embraces until the Oligocene appears, when they settle the carbonated rocks, with some terrigenous content (Paso Real Formation), those which do not appear in the area (García et al. 1987).

Lying discordantly on Paso Real Rocks, the carbonated reef environment constitutes the main nucleus of the emerged surface of the Peninsula; and the rocks of more areal distribution (Vedado Formations) appear, and terrigenous of continental alluvial origin, appears toward the east of the region (Guane Formation). The age of both Formations is considered among the Superior Pliocene until the Inferior Pleistocene. Rocks of Half Pleistocene to Superior age (Jaimanitas Formation), represented by biodetritus and coralline calcareous reef, lie on the Vedado Formation, occupying the first marine terrace. With Half Pleistocene age, possibly extending to the Superior, the biocalcarenite and oolitics rocks (Cayo Piedras Formation) were found. Crowning the cut finds Holocene deposits of diverse genesis (marines, biogenetics, eluvials, and secondary forms of carbonate of calcium; Denis et al. 2005).

Its main morphological feature is the prevalence of naked and partially covered karrenfield, associated with small covering dolines almost always of red soil and dolines, mainly of dissolution and collapse, to those that associate caves in a majority of phreatic origin, marine or phreatic-marine, with great influence of vadose processes and very wide development of sufosive-karstics processes, without the presence of superficial drainage. The area is an active karstified plain, where several characteristic types of relief are appreciated. Therefore, much emerged as being submerged, as it is karst, assisting the engineer geologic classification (Skwaletski and Iturralde-Vinent 1971), and keeping in mind the stratification, a homogeneous massif for the composition of carbonated sequences, of deep drainage, for their age it can be classified as young, and for their relationship with the surface, naked and partially covered (Denis et al. 2005).

10.3 Methodology

Mapping the intrinsic vulnerability of underground waters was carried out following the multi-parameter approach EPIK (Doerfliger et al. 1999; Tripet et al. 2000), which considers four factors to which are assigned a specific weight and a value for each one of their criteria. The factors are:

- **E** (epikarst develop) is defined indirectly for morphological analysis by means of photo interpretation to scale 1:33,600, topographical maps (1:25,000), geologic maps (1:50,000) and field work.
- **P** (covered protector) is characterized by means of the soil depth, which is closely related with the time of residence of the pollutants (Doerfliger et al. 1999). The used database was the Soil Map (1:25,000) and the geologic map (1:50,000) this last one, for the presence of quaternary deposits that overlie the aquifers.
- **I** (infiltration conditions): this parameter represents the concentrated or diffuse conditions of the infiltration waters (duality of the recharge). However, it does not include the recharge in terms of quantity or localization (Doerfliger et al. 1999). Its determination is carried out by means of the analysis of topographical maps and fieldwork.
- **K** (develop of karst network): this factor defines the organization degrade of the aquifer, playing an important role in the flow velocity and, consequently in the vulnerability. The speleological exploration, geophysical methods and tracer test, as well as the analyses of temporary series are the main sources of information for the characterization of this factor.

The quantification and evaluation of the vulnerability is carried out by means of the application of Eq. 10.1, being obtained with the help of the SIG, the protection factor F*pi*, being defined as the vulnerability classes such as (Table 10.1):

$$F_{pi} = \alpha E_i + P_i + \gamma K_i \qquad (10.1)$$

Where: $i = 1,.., n$ is the value of the cell of the gid; E_i, P_i, I_i, K_i = weight assumed in the cell i; α, β, γ = relative weight of the attribute (constant for

Table 10.1 Vulnerability class and protection factor (Doerfliger et al. 1999)

Vulnerability class	Protection factor
Very high	$F \leq 19$
High	$19 < F \leq 25$
Moderate	$F > 25$
Low	$F > 25$; $P = P_4$; $I = I_{3,4}$

each attribute); F_{pi} = factor of protection of the cell i, the smallest protection factor calculated for any cell *i* shows the biggest vulnerability in the aquifer.

In all the parametric techniques the subjectivity in the selection of rating and weights related to the parameters is inevitable. This subjectivity affects notability in the final result. In such a sense, sensitivity analysis provides valuable information on the influence of rating values and weights assigned to each parameter and help the analyst to judge the significance of subjective elements (Gogu and Dassargues 2000). For the application, a concept of unique subarea conditions is used. "*Map removal*" and "*Single parameter*" sensitivity analyses were performed. Sensitivity of removing one or more maps can be expressed using Eq. 10.2 (Lodwik et al 1990):

$$S_{Xi} = \left| \frac{V_i}{N} - \frac{V_{Xi}}{n} \right| \tag{10.2}$$

Where: S_{Xi}: sensitivity (for the ith unique condition subarea) associated with the removal of one map; V_i: Vulnerability computed using Eq. 10.1 on the ith subarea; V_{Xi}: Vulnerability of the ith subarea without considering parameter X; N an n: number of maps used to compute R_i and $R_{Xi,}$, respectively.

The variation index measures the effect of the removal of each parameter. The magnitude of the variation created by removal of one parameter was computed using Eq. 10.3 (Gogu and Dassargues 2000):

$$RX_i = \frac{V_i - V_{Xi}}{V_i} \cdot 100 \tag{10.3}$$

Each parameter contributes with an effective weight to the vulnerability. The single parameter sensitivity analysis compares the effective weightings of the each input parameter with the empirical weighting assigned by the analytical model (Napolitano and Fabbri 1996). The effective weighting can be calculated for each subarea using Eq. 10.4:

$$W_{Xi} = \frac{X_{Vi} \cdot X_{Wi}}{V_i} \cdot 100 \tag{10.4}$$

Where X_{Vi} and X_{Wi} are, respectively, the raring and the weighting values for the parameter X assigned in the subarea i. V_i is the recharge computed in Eq. 10.1.

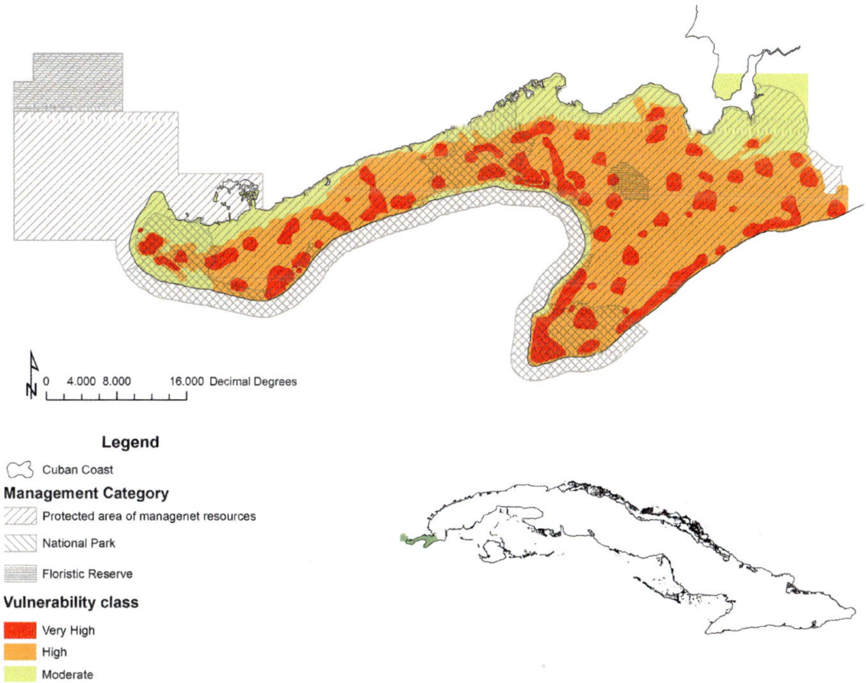

Fig. 10.1 Vulnerability maps of Guanahacabibes National Park

10.4 Result and Discussions

In the area, very high and high vulnerability occupies approximately 80 % of the area (Fig. 10.1). These vulnerability classes were found in carbonate karst aquifers, with different karstification degrees. Very high vulnerability areas occupy 16.72 % of the area and it is associated with extensive karrenfield, dolines and cenotes, directly connected to the groundwater. High vulnerability is the mostly extended in the territories (63.13 %). In this area karst development is moderate, principally associate with fissured and dissolution features.

In the area of Moderate vulnerability (20.15 %) is found karst development, which is incipient. In this area alluvial deposits are found.

To this resultant map, sensitivity analysis is applied to define the influence of each factor and calculate the real weighting in the area. A result of map removal sensitivity analysis is shown in Table 10.2.

Gogu and Dassargues (2000) in small karst aquifers in southern Belgium found that E factor is more sensible that I factor. However, the relative influence on the

Table 10.2 Statistics of map removal sensitivity analysis

Parameter	Average	Minimum	Maximum	Range	Standard deviation
E	296	2	3.91	1.91	0.79
P	2.79	2.08	3.75	1.67	0.5
I	3.27	2.67	3.91	1.24	0.48
K	2.83	2	3.91	1.91	0.74

Table 10.3 Statistics of single parameter sensitivity analysis

			Effective weighting (%)					
Parameter	Empirical weighting	Empirical weighting (%)	Average	Minimum	Maximum	Range	Standard deviation	Real weighting
E	3	33.33	29.41	15	42.86	2,786	9.27	2.65
P	1	11.11	7.98	4	12.5	85	2.93	0.72
I	3	33.33	46.49	33.33	60	2,667	9.42	4.18
K	2	22.22	16.12	9.09	31.58	2,249	6.65	1.45

final vulnerability map in Guanacahabibes area is I > E > K > P. This shows that the sensitivity analysis varied with the study area. The EPIK method is very sensible to remove factor I, due to the criteria that define the duality of recharge. The other factor shows similar sensitivity to removing, with a range of 0.17 between them. The variation indexes of each parameter are always positive, which suggests that the exclusion of any parameter reduces the recharge, which, therefore, increases the computation.

The single parameter sensitivity analysis shows some deviations between the empirical and real weighting (Table 10.3). Parameters E, P and K are overvalued. The real importance of infiltration is major in this particular analysis, which was assigned in the original method. In the analysis, it is deducted that the result of vulnerability index is dominated by factor I with an average of 4.18 (46.49 %) against the empirical weighting of 3 (33.33 %). The spatial distribution and the relationship between four parameters are factors, which determine the final real or effective weightings.

As the single parameter sensitivity analysis defines the real weighting of each factor in the area, the results are used for the assessment vulnerability map. Two resultant maps, one with the average of real weighting and other with the real weighting in each cell are shown in Fig. 10.2.

The results show a notable difference in the final maps, especially when the average value of the analysis is used. As a point Farfán et al. (2010) defines an ideal weight by means that the average is slightly effective and shows similar and not more efficient results. In this case the use of average reduces the very high vulnerability areas in sectors when karst landforms as dolines and cenotes with

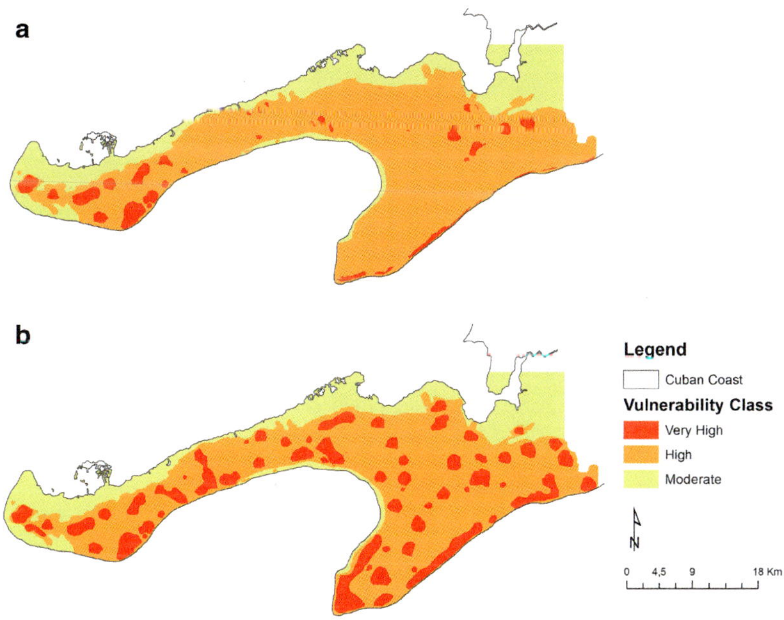

Fig. 10.2 (**a**) Vulnerability map using the average of effective weighting. (**b**) Vulnerability map using effective weighting in each cell

direct relationship with the water table are found. However, the use of the real value in each cell shows a best approach, but shows too little difference in the high vulnerability areas.

10.5 Conclusions

The EPIK methods are applied from the assessment vulnerability map of Guanacahabibes Peninsula, due to easy realization and taking into account the accessible database of the area. The first result shows that more than 80 % of the area has very high and high vulnerability. Low and very low class are completely absent, due to the hydrogeological units and especially karst development.

For determining the susceptibility of each parameter, a sensibility analysis is realized. The analyses show that the infiltration parameter has major influence in the final map, in contrast to protection cover with a slow influence.

The vulnerability map of the Guanahacabibes National Park has provided useful information for planning activities at a regional level. The next step of the research is to acquire data on the chemical and physical parameters of the studied area and on the land use in order to realize a map of the hydrological risk of the area.

References

Barrocu G, Muzzu M, Uras G (2007) Hydrogeology and vulnerability map (Epik method) of the "Supramonte" karstic systems, north-central Sandinia. Environ Geol 51:701–706

Cucchi F, Franceschini G, Zini L, Aurighi M (2008) Intrinsic vulnerability assessment of Sette Comuni Plateau aquifer (Veneto Region, Italy). J Environ Manage 88:984–994

Denis R, Díaz C, Rosa CR, Hernández PL (2005) Características geológicas y geomorfológicas de la Península de Guanahacabibes. Cong. 65 Aniv. De la SEC. Viñales. Pinar del Río

Doerfliger N, Jeannin P, Zwahlen F (1999) Water vulnerability assessment in karst environments: a new method of defining protection areas using a multi-attribute approach and GIS tools (EPIK method). Environ Geol 39(2):165–176

Farfán H, Rodríguez Y, Días JE (2009) Vulnerabilidad a la contaminación de las aguas subterráneas de la cuenca del Arroyo de Santo Tomás, Viñales, Cuba. MAPPING No. 137: 60–64

Farfán H, Corvea JL, de Bustamante I (2010) Sensitivity analysis of A.P.L.I.S. method to compute spatial variability of karst aquifers recharges at the National Park of Viñales, Cuba. In: Andreo B, Carrasco F, Duran JJ, LaMoreaux JW (eds) Advances in research in karst media. Springer, Heidelberg, pp 19–25

García, A, Victorero C, Hernández S (1987) Apunte geológico sobre la parte occidental de la Península de Guanahacabibes. Instituto Superior Pedagógico, Pinar del Río

Gogu RC, Dassargues A (2000) Sensitivity analysis for the EPIK method of the vulnerability assessment in a small karstic aquifer, southern Belgium. Hydrogeol J 8:337–345

Goldscheider N (2002) Hydrogeology and vulnerability of karst systems, examples from the Northern Alps and Swabian Alb. PhD thesis, Univ. of Karlsruhe, p 236

Lodwik WA, Monson W, Svoboda L (1990) Attribute error and sensitivity analysis of map operations in geographical information systems-suitability analysis. Int J Geog Inf Sist 4:413–428

Napolitano P, Fabbri AG (1996) Single-parameter sensitivity analysis for aquifer vulnerability assessment using DRASTIC and SINTACS. HydroGIS 96: application of Geographic Information Systems in Hydrology and Water Resources Management (Proceedings of the Vienna Conference, April 1996). IAHS Publ. no. 235. pp 559–566

Petrič M, Šebela S (2004) Vulnerability mapping in the recharge area of the Korentan spring, Slovenia. Acta Carsologica 33(2):151–168, Ljubljana

Ravbar N (2007) The protection of karst waters: a comprehensive Slovene approach to vulnerability and contamination risk mapping. Karst Research Institute at ZRC SAZU, Ljubljana. Carsologica 6, p 254

Ravbar N, Goldscheider N (2009) Comparative application of four methods of groundwater vulnerability mapping in a Slovene karst catchment. Hydrogeol J 17:725–733

Skwaletski EN, Iturralde-Vinent M (1971) Estudio ingeniero-geológico del carso cubano. Serie Espeleológica y Carsologica 31:58

Tripet J-P, Doerfliguer N, Zwahlen F, Delporte C (2000) Vulnerability mapping in karst areas and its uses in Switzerland. Acta Carsologica 29(1):163–171, Ljubljana

Zwahlen F (2004) Vulnerability and risk mapping for the protection of carbonate (karstic) aquifers. Final report COST action 620. – European Commission, Directorate-General for Research, Brüssel, Luxemburg, p 297

Chapter 11
First Outcomes in the Application of PaPRIKa Method to Assess Aquifer Vulnerability in Tropical Karst Mountain: Santo Tomás Watershed: Viñales National Park, Cuba

H. Farfán-González and V. Plagnes

Abstract For Protected Areas, the vulnerability of groundwater maps are highly desired as management tools, because they help in delimiting protection zones, in classifying their importance, and thus show where most management attention is required. From karst terrains, due to their high hydrogeological complexity, several methods that take into account their intrinsic properties have been developed. Recently, and as a logical evolution of EPIK, PaPRIKa method has been developed from the definition of protection criteria in karst aquifers in France. In the present paper, the PaPRIKa method is applied to aquifers in a tropical karst mountain – Santo Tomás watershed – west of Viñales National Park, Cuba. Due to the flexibility of this multi-criteria method, four resultant maps based on different weighting systems, defined different scenarios of vulnerability. In the karst reservoir the variability is more noticeable; however, the statistical comparison analysis shows that high and very high vulnerability are common. The variation in the result is more evident in the moderate and low vulnerability. For the validation of each vulnerability resultant maps and to determine the best approach, a multi-tracer test is necessary.

Keywords Protected areas • Groundwater vulnerability mapping • Tropical karst • PaPRIKa method • Cuba

H. Farfán-González (✉)
PELA GeoEnvironmental, Inc., 1009 23rd Avenue, 35401, Tuscaloosa, Alabama, USA
e-mail: jlamoreaux@pela.com

V. Plagnes
UMR Sisyphe Univ. Pierre et Marie Curie (UPMC), Paris, France
e-mail: valerie.plagnes@upmc.fr

H. Farfán González et al. (eds.), *Management of Water Resources in Protected Areas*, Environmental Earth Sciences, DOI 10.1007/978-3-642-16330-2_11,

11.1 Introduction

Initially, the term "vulnerability of aquifers" was used without any definition, but it does refer to susceptibility of the aquifers affected by anthropogenic contamination (Foster et al. 2002). Although the vulnerability of aquifers is very difficult to define as it cannot be measured or directly obtained in the field (Vrba and Zaporozec 1994), nowadays, their mapping is a very important management tool.

In karst aquifers the problems of assessing groundwater vulnerability are more complex, due to their high heterogeneity and anisotropy as a direct consequence of the duality of the hydrological processes. Other characteristics are good indicators of the susceptibility to pollution as: high distribution of permeability; the strong connection between the superficial and subterranean drainage subsystems; the variability of response to stimulus depending on the inertial characteristics of the system; the temporal inversion of the hydraulic gradient and the presence, in many cases, of a discontinue surface, the presence of epikarst, with functions of storage and regulation of flow in the unsaturated zone; and the resident time is frequently very short and with a limited interaction with the media, where a good part of the attenuation processes is not sufficiently effective (Zwahlen 2004; Molerio 2002; Farfán et al. 2009; Ravbar 2007; etc.).

For these reasons, in recent times, several methods that take into account these specific properties for best estimation to groundwater vulnerability mapping in karst territories have been developed. The most important results are achieved in the COST Action 620, where the bases for vulnerability and risk mapping to European scale are resumed (Zwahlen 2004). Most recently, and as a logical evolution of EPIK, the PaPRIKa method has been developed for the definition of protection criteria in France karst aquifers (Dörfliger and Plagnes 2009; Plagnes et al. 2010).

Especially for protected karst areas, mapping vulnerability is a very useful tool because they help in delimiting protection zones, classify their importance, and thus show where most management attention is required (William 2008). In the present paper, the PaPRIKa method is applied to aquifers in a tropical karst mountain – Santo Tomás watershed – west of Viñales National Park, Cuba. Due to the flexibility of the method, four resultant maps based on different weighting systems, defined different scenarios of vulnerability.

11.2 Description of the Site

The Santo Tomás watershed is part of the Sierra de los Órganos, in western Cuba. It is characterized by a typical tropical karst developed in Jurassic – Cretaceous carbonate rocks. The typical landscape presents low karst denudated mogote-type mountains, marginal poljes and dolines, and a development of extensive karrenfield and caves.

Geologically, three deposits are recognized. One terrigenous, corresponding to the oldest deposits in a Jurassic delta complex formed by sandstones, clays, schists. Unconformably, above these deposits, the massive or grossly stratified limestones are present (Upper Oxfordian – Lower Tithonian), and another terrigenous with olistostromic character (Middle Eocene), produced by turbidites associated with overthrust of the complex at Cretaceous Arc over Guaniguanico deposits. An Upper Pliocene – Upper Pleistocene marine transgression determined the formation of an alluvial plain, and deposition of the terrigenous material. Further alluvial and eluvial deposits of recent age complete the stratigraphic setting of the area (Farfán et al. 2010).

Sierra de Quemado is an anticline fold with a charnel of 30° of direction and with an axial plane practically vertical in their SW position that buries itself in a NE direction. It is a fold in which axis has been deformed by the differential movement of the principal tectonic event in the geological history of the region, the emplacement of tectonic mantel from SE, and maybe due to the posterior development of Pinar Fault, represented then, as a structure of second order (Morales 2010).

From the groundwater viewpoint the area belongs to the Cuyaguateje basin, and consists on five sub-watersheds of the asymmetric dendritic type: Santo Tomàs, Peñate, el Bolo, la Tierra, and los Cerritos. The waters from the five sub-watersheds join in the Sierra de Quemado to give origin to the Gran Caverna de Santo Tomás (46 km), the longest one in Cuba, composed of seven cave levels of galleries, and with a potential of some 70 km of development within the subterranean system of Sierra de Quemado, as demonstrated by tracer tests (Molerio 2004). The Gran Caverna de Santo Tomás has a hydrologically active level (the lower one), two seasonal levels: occasional or episodic functioning, and four fossil levels. Several permanent lakes, mostly fed by autochtonous water, are present in the seasonal levels of the system. The main discharge occurs with the river Santo Tomás, integrated by allogenic inputs (Farfán et al. 2010).

11.3 Methodology

PaPRIKa is a multi-criteria method based on index and weighting system. The name is the acronym of: "**P**rotection of karst **a**quifers based on their **P**rotection; **R**eservoir; **I**nfiltration and **Ka**rstification degree". This method consists of resource and source mapping. In this paper only the vulnerability of the resource method are made.

The four attributes of PaPRIKa were defined regarding the structure (P and R criteria) and hydrogeological behavior (I and Ka criteria) of karst aquifers (Plagnes et al. 2010). The weighting of four criteria is equal to 1. The system of weighting is very flexible, but the total sum of the weight at the structure is always equal to 0.4 and the total weight at the hydrogeological functioning is equal to 0.6, to put more emphasis on the special behavior of this type of aquifer. For the swallow hole basin, only P and I are evaluated, and always with the same weight of 0.5 from each criteria. The degree of vulnerability is quantified by an index that is the sum of the products of each parameter weight in this rating. The distribution of the vulnerability index within five intervals of the same magnitude allows the specification of the different vulnerability classes of the PaPRIKa method. The four Eqs. 11.1, 11.2, 11.3 and 11.4 used to assess vulnerability in the karst reservoir in present studies are:

$$V = 0.2P + 0.2R + 0.4I + 0.2Ka \quad (11.1)$$

$$V = 0.25P + 0.25R + 0.25I + 0.25Ka \quad (11.2)$$

$$V = 0.2P + 0.1R + 0.4I + 0.3Ka \quad (11.3)$$

$$V = 0.2P + 0.2R + 0.5I + 0.1Ka \quad (11.4)$$

The results of the four equations have been compared by statistical methods. The relation between the factors and each index of vulnerability can be obtained statistically using the correlation coefficients (Vías et al. 2010). In the present study, the determination coefficient (square of Pearson's correlation coefficient) was used to measure the variability in the distribution of each PaPRIKa index resultant, respect ting the distribution of each criteria score.

11.4 Results and Discussions

The vulnerability maps using the PaPRIKa index (Fig. 11.1) have been made by overlaying the four criteria with different weigh systems according to the original method (Dörfliger and Plagnes 2009), by a GIS analysis (ArcGIS 9.3). For the Protection (P) shape the geological map of the area to scale 1:50,000, the soil map (1:25,000) and the water table map (1:25,000)-this last one used to determine the depth of unsaturated zones – were used. Reservoir (R) criteria is estimated by the geological map, and the Infiltration (I), was determined using a digital elevation model (DEM) of the region (1:25,000) reclassified in the corresponding intervals, the interpretation of the topographical map (1:25,000) and aerial photos (1:33,600). Karstification (Ka) criteria are mapped by speleological topography and the result of the tracer test (Molerio 2004).

The applications of the PaPRIKa method show the spatial variability of the physical proprieties field in the Sierra de Quemado reservoir. The areas of alochtonous flow (annex basin), when the superficial runoff are developed, is considered as high and very high vulnerability, because the drainage enters directly to the reservoir via swallow hole, and the same criteria and weighting from the calculus is used. In the karst reservoir the spatial variability in each application is more noticeable, because the four criteria are used, with different weighting systems. Very low vulnerability only differs in the version 3 from the application, and it is the one that has larger spatial variability and homogeneous distribution of the vulnerability. Several areas of very high, high and moderate vulnerability are common for the four application; however, the second version that was carried out with the same weight for each approach presents a remarkable prevalence of the high vulnerability, where the approach Ka defines the areas of very high vulnerability.

The zoning of the vulnerability classes by PaPRIKa in each different scenario is defined by the criteria regarding the hydrogeological behavior. The statistical analysis performed to the PaPRIKa criteria reassures the importance of Factor I showing

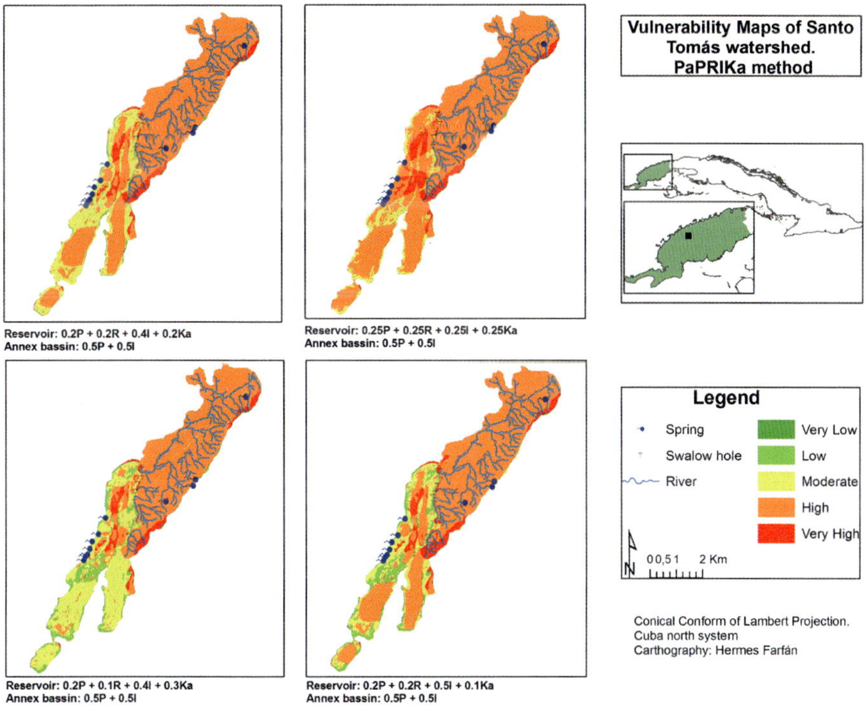

Fig. 11.1 Result of the application of PaPRIKa method for vulnerability of karst aquifers in the Santo Tomás watershed, Viñales National Park, Cuba

their high influence on each final map. However, in PaPRIKa v2, these criteria show the lowest correlation, motive by the weight system used in the calculus.

The influence of the other criteria is very low without remarkable influence on the final maps (Table 11.1).

This aspect is motivated by the strong karstification where the most important karst landforms are dolines and karrenfield. In these areas, when the soil is incipient and karst landforms are remarkably developed, the vulnerability is generally Very High and High. The criteria of structure do not show a notable influence on the result. This aspect is motivated by the homogeneity of the reservoir and the unsaturated zone, although in the P criteria, the thickness is perhaps very influent in some specific areas.

The same analysis is applied to define the spatial relationship between the different vulnerability maps. The coefficients of determination show a strong spatial correlation between the applications, except from PaPRIKa v2 and PaPRIKa v4 ($r^2 = 0{,}191$), due to the notable difference in the weight systems. From PaPRIKa v2, the same weight is used and the spatial variability of the vulnerability is more limited to the rating values.

The statistical and visual inspection allows only the defining the spatial relationship between the criteria used in the confection of each vulnerability map and the final maps.

Table 11.1 Statistical correlation (coefficient of determination) among vulnerability criteria and PaPRIKa index

	Vulnerability index			
Criteria	PaPRIKa_v1	PaPRIKa_v2	PaPRIKa_v3	PaPRIKa_v4
P	0.0449	0.145	0.038	0.024
R	0.0247	0.010	0.044	0.025
I	0.791	0.541	0.740	0.887
Ka	0.099	0.217	0.142	0.033

11.5 Conclusions

The PaPRIKa method is applied for the first time in the aquifers of a tropical karst mountain in Cuba. In each approach, the differentiation of the vulnerability is conditioned by the infiltration criterion (I) that is related to the hydrogeological behavior. Although for visual inspection the different results show differences, the statistical correlation between the different approaches shows a high spatial correlation, except from the PaPRIKa v2 and PaPRIKa v4. This is sustained by the notable difference in the weight systems. However, the results of the different weight criteria show the spatial variability of the vulnerability in the reservoir. Future analyses in the area should be directed to the validation by a multi-tracer test in several points and different hydrological conditions for the definition of a response plan against any contamination events.

Acknowledgments The results were achieved due to the activities supported by the Viñales National Park Management Plan, especially the investigation programs: "*Hydrogeology and vulnerability of karst aquifers in the Viñales National Park*".

References

Döerfliger N, Plagnes V (2009) Cartographie de la vulnérabilité intrinsèque des aquifères karstiques. Guide méthodologique de la méthode PaPRIKa. Rapport BRGM RP-57527-FR. p 100

Farfán H, Diaz C, Corvea JL, et al (2009) Comparación estadística de diferentes metodologías para la cartografía de la vulnerabilidad de acuíferos cársicos. 3ra. Convención Cubana de Ciencias de la Tierra. GEOCIENCIAS 2009. 16–20 de Marzo. La Habana, Cuba: ISSN: 978-959-7139-83-6

Farfán H, Diaz C, Parise M (2010) Scenarios of groundwater pollution in a karst watershed: a case study in the Pinar del Río province at Cuba. In: Andreo B, Carrasco F, Duran JJ, LaMoreaux JW (eds) Advances in research in karst media. Springer, Berlin/Heidelberg, pp 289–295

Foster S, Hirata R, Gomes D et al (2002) Protección de la calidad del agua subterránea. Guía para empresas de agua, autoridades municipales y agencias ambientales. Groundwater Management Advisory Team. WHO-PAHO-CEPIS y UNESCO-ROSTLAC-PHI. Banco Mundial Washington DC. p 127

Molerio LF (2002) Indicadores de vulnerabilidad de acuíferos cársicos. Subprograma XVII: Gestión de recursos hídricos. II Taller de Vulnerabilidad de Acuíferos. p 8

Molerio LF (2004) El enlace absorción-descarga de la Gran Caverna de Santo Tomás: evidencias derivadas de un ensayo con trazadores artificiales. Ing Hidr y Ambiental XXV(3):22–26, La Habana

Morales GEC (2010) Estructura geológica y su vinculación con las morfoestructuras del área de la cuenca del arroyo Santo Tomás y la Sierra de Santo Tomás – Quemado. Thesis Ing. Geol. Univ. Pinar del Río, Cuba, p 108 (unpublished)

Plagnes V, Kavouri K, Huneau F (2010) PaPRIKa, the French multicriteria method for mapping the intrinsic vulnerability of karst water resource and source-two examples (Pyrenees, Normandy). In: Andreo B, Carrasco F, Duran JJ, LaMoreaux JW (eds) Advances in research in karst media. Springer, Berlin/Heidelberg, pp 323–328

Ravbar N (2007) The protection of karst waters: a comprehensive Slovene approach to vulnerability and contamination risk mapping, vol 1. Karst Research Institute at ZRC SAZU, Ljubljana, p 254, Carsologica

Vías J, Andreo B, Ravbar N, Hötzl H (2010) Mapping the vulnerability of groundwater to the contamination of four carbonate aquifers in Europe. J Environ Manage 91:1500–1510

Vrba J, Zaporozec A (1994) Guidebook on mapping groundwater vulnerability. Int Contrib Hydrogeol 16:131

William P (2008) World heritage cave and karst. A thematic study. IUCN-WCPA. World Heritage Studies. No. 2, p 57

Zwahlen F (2004) Vulnerability and risk mapping for the protection of carbonate (Karstic) aquifers. Final report COST action 620. – European Commission, Directorate-General for Research, Brüssel, Luxemburg, p 297

Chapter 12
Assessment of Groundwater Protection Zones at the Viñales National Park, Cuba

H. Farfán, J.L. Corvea, and I. de Bustamante Gutiérrez

Abstract Because of the detection of several pollution hazards in the park area and its surroundings, groundwater protection in the Viñales National Park has been defined as a priority in its Management Plan for the next 5 years (2009–2013). These hazards directly affect the surface waters, and the karst nature of the territory allows the contaminants to enter directly to the aquifers without any or almost negligible self-depuration processes. In turn, these can affect very large areas in a short time. In this paper several results of precedent studies related to the vulnerability, hazards, risk, and the mapping of the spatial variability of natural recharge are integrated in a GIS analysis. The methodology is based on Jimenez-Madrid et al. (2010) slightly modified in this contribution. The analysis of hydrographs, chemographs, mathematical models and tracer tests become necessary for the definition of a response plan against any contamination events and for model validation, together with the incorporation of data related to the energetic responses from the aquifer system to external forcing.

Keywords Pollution hazards • Groundwater protection • GIS analysis • Cuba

12.1 Introduction

The protection of groundwater requires a holistic knowledge based on the exhaustive perception of the behavior of aquifers. In the case of karst aquifers, the problem of the definition of the protection perimeters of water resources is even more complex, due to

H. Farfán (✉) • J.L. Corvea
PELA GeoEnvironmental, Inc, 1009 23rd Avenue, 35401, Tuscaloosa, Alabama, USA
e-mail: jlamoreaux@pela.com

I. de Bustamante Gutiérrez
Instituto IMDEA agua, c/Punto Net, n°4, Edificio Zye, 2ª Planta, E-28805 Alcalá de Henares, Madrid, Spain
e-mail: irene.bustamante@imdea.org

H. Farfán González et al. (eds.), *Management of Water Resources in Protected Areas*, Environmental Earth Sciences, DOI 10.1007/978-3-642-16330-2_12,

the duality (diffuse and concentrate flow) of the organization of runoff due to the heterogeneity and anisotropy of the flow domains. Even more, if it is treated or defined as a binary (hybrid) system, allogenic waters enter the system in a concentrated flow path through swallow hole, ponors or caves and are mixed with the autochthonous waters percolating within the limits of the karst system. Because of this hydrological complexity, the definition of the adequate approach for the protection perimeters of water resources cannot be defined by conventional techniques but by new approaches accounting for the specific properties of karst systems.

The Sierra de los Órganos is one of the most singular scenarios of Cuban karst with an exceptional development of the cone and tower karst (*kegelkarst*) in Jurassic-Cretaceous rocks. In this mountainous massif where the tower karst (locally named mogotes) is broadly represented, several binary (hybrid) karst systems coexist. Because of their geomorphologic, biological and aesthetic values, several protected areas of different categories have been declared. The most representative of them is the Viñales National Park, with notable scientific, historical, natural and aesthetic values.

Farfán (2010) has summarized the state of knowledge of the Viñales National Park hydrogeology. The integration of several hydrodynamical, hydrochemical, tracer tests, self-purification capability of underground waters, vulnerability, the inventory of the main polluting sources and the chemical quality in the Santo Tomás watersheds, protection zones have been defined and validated by Farfán et al. (2010c).

However, the necessity to extend this approach to the whole Viñales National Park is one of the priorities of the protected area.

This contribution integrates the results of the previous hydrogeological investigations in a map that shows the protection zones of Viñales National Park (Aldana et al. 2007; Díaz and Farfán 2007; Farfán 2010; Farfán et al. 2005, 2009, 2010a; Molerio 1995, 2004; Molerio and Flores 2003, etc.), in an attempt to define the appropriate strategies of water management and land use. Some consideration about the convenience of incorporating to the Jiménez-Madrid et al. (2010) data method related to the energetic response to the aquifer systems to external forcing is discussed. The complementary analysis of hydrographs, chemographs, simulation models and tracer test is necessary for the proper definition of a response plan against any contamination event and for the model validation.

12.2 Description of the Site

Viñales National Park is a protected area that comprises a remarkable number of natural, scientific, historical and aesthetic values. Two of them are the Gran Caverna de Santo Tomás (National Monument) and the Viñales Valley, declared by the UNESCO "Cultural Heritage" in 1999.

It is located in the Center-oriental region of Sierra de los Órganos, where one of the most spectacular exponents of residual tropical karst of mountains in Jurassic-Cretaceous carbonated rock is development. The karst is classified as *conical and towers in complexly folded and failed heights, constituted by potent strata of carbonated and not carbonated rocks,* with complex alpine tectonic features (Núñez et al. 1965) that are manifested with the presence of tectonic nappes.

The geomorphology of the region in the strictest sense, consists of a genetic complex of the structural relief and fluvio-karst of calcareous mountains and mogotes, where the positive forms of landforms have polihydraulics contours. They could be petrogenic, but basically they reflect the regional fissure pattern and present the disposition of lengthened partitions and are parallel to the direction of the layers (Gutiérrez 1995).

The relief has been classified as low karstic-denudative mountains of structure-karstified type, in those that stand in particular landforms as: the isolated mogotes (hum), chain mogotics (mountains), poljes and dolines (marginal and interiors); extensive karrenfields in their most varied dimensions and morphologies, as well as a huge cave development (Díaz 1999).

In the Sierra of the Organs, in general sense, the direction of underground flow takes place as a consequence of the transfluence and transcurrents of the superficial, and partially underground drainage networks, and also as a result of the discharge of an autochthonous flow bag that constitutes a peculiar modality of drainage. The recharge, movement and discharges of waters are made through the karst landforms, many times connected in united circuits. The most important conduction landforms are direct cavities of allogenic and autochthonous flow, while the emission landforms are mainly of allogenic flows that generally discharge on the fluvial valleys (Molerio et al. 1997).

On the other hand, in this area, they are the biggest Cave Systems of Cuba and probably of The Antilles, with a rich fauna of vertebrate fossils and a remarkable number of archaeological evidences. These are the cases of the Gran Caverna de Santo Tomás (46 km), Palmarito-Novillo-Pan de Azúcar Cave System (48 km), the Guasasa Underground System (15 km), the San Vicente System (6 km), among others.

12.3 Methodology

The methodology of Jiménez-Madrid et al. (2010) for the definition of the groundwater protection zones for human consumption was slightly modified in this paper. The methodology consists of three stages. Two of them, as shown below were applied. The third was not taken into account because the recharge areas of most springs are not well defined.

- **First stage**: Zoning is obtained by means of risk assessment. For the evaluation of the intrinsic vulnerability the PC method was applied. PC method is a simplification of COP method, where only C (P in this approach) and O factor are valued due to the different levels of knowledge of the local aquifers in the area. This approach is more extended than the Duality Method (Nguyet and Goldsheider 2006). From pressures (hazards) the classification of Nguyet and Goldsheider (2006) is assumed. As a result of this step, the first zoning has been defined.
- **Second stage**: Evaluation of aquifer recharge of the area, based on the APLIS method (Farfán et al. 2010b), developed specifically for carbonated aquifers (Andreo et al. 2008; Marín and Vias 2010). A second zoning is obtained by

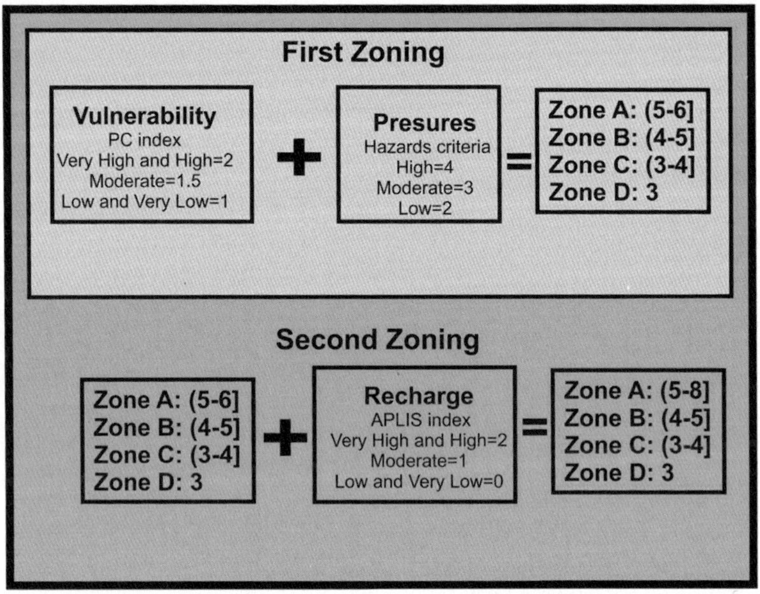

Fig. 12.1 Used methodology. Modified from Jimenez-Madrid et al. (2010)

means of the overlaying of the result obtained in the first stage and the mapping of the spatial distribution of recharge Fig. 12.1.

The results of each stage are grouped into four zones following the protection approaches (Jiménez-Madrid et al. 2010):

- ZONE A: Safeguard zone with heavy restrictions.
- ZONE B: Safeguard zone with moderate restrictions.
- ZONE C: Safeguard zone of future prevention.
- ZONE D: Safeguard zone not established.

12.4 Result and Discussions

12.4.1 Groundwater Protection Zones in Viñales National Park

Table 12.1 shows the area occupied by each protection zone. The vulnerability map (Farfán 2010) shows that all the karst aquifers exhibit a very high and high vulnerability while the sectors of karst valley and territories of terrigenous rocks shows low and very low vulnerability. Pressures are represented by high hazards in the sector where urban development and punctual sectors when untreatable wastewaters exist. Several areas with agricultural use and diffuse contamination, specifically by nitrates, are classified as medium or moderate pressures. In the area, extensive and intensive agricultures exist, and approximately 1 ton/ha/year of

Table 12.1 Percent of the territory occupied by each one of the zoning stages, for the delimitation of the protection zone

	Protection zones			
	Heavy restrictions	Moderate restrictions	Future prevention	Not established
First zoning (risk)	0.10	5.95	51.71	42.24
Second zoning	9.66	41.00	17.95	31.39

chemical fertilizer (NPK complete formula) is used in the soil dedicated to tobacco crops (Farfán et al. 2010). The rest of the area is defined as low pressure because no anthropogenic activities of potential contamination are developed.

The combination of these two maps gives the first zonation. Zones of future prevention and unestablished zones prevail (94.95 % of the territory). In this first zoning (equivalent to a simplified risk map) heavy restriction is very scarce and moderate restrictions are found principally in sectors where high vulnerability and moderate pressures spatially coexist.

The evaluation of the spatial distribution of natural recharge in the area (Farfán et al. 2010b) is superimposed to the first zoning to obtain a final result. A notable difference between the two zonings is observed. For example, Zone A (heavy restrictions) is almost absent in the first approach, while in the second it occupies an area equivalent to 10 % of the territory. This is remarkable in Fig. 12.2. The smallest variation is observed in Zone D, because it is associated with low productive aquifers.

Heavy restrictions are required for 9.10 % of Viñales National Park. These sectors are mainly associated with zones where moderate and future prevention zones and moderate and very high-high recharge converge, respectively.

They are associated with the development of the naked karst with extensive and very well developed karrenfields and with a great number of dolines (mainly collapsed). In some cases, they are linked to the active or seasonal levels of groundwater development. In these areas, epikarst is remarkably developed, while the soil is little developed or nonexistent. These are also the main points of absorption of the allogenic recharge, constituted by the swallow hole of absorbent cavities of allochtonous flow that conform each underground system. Some sectors of the Viñales Valley, due to the scarce development of the soil, can also be areas of scarce protection.

Moderate restrictions occupy an area of 41 %, and it is associated essentially to zones in Zone C and moderate recharge. In these areas, karst development is not so representative, and only some dissolution features are present. On other hand, in some areas of the poljes and adjacent valleys, where the agricultural development is notable and the vulnerability is from moderate to high, moderate restrictions are present.

For the zone of Future prevention (17.95 %), it is very different in the carbonate-terrigenous aquifers, and in some areas of the karst valleys and marginal poljes when deeper soils are developed. In these areas, no pressures are observed. Areas of stream development in non-carbonated rocks show similar zoning criteria. In the unestablished zone (31.39 %), it is associated with aquifers with scarce resources composed of shales and silts.

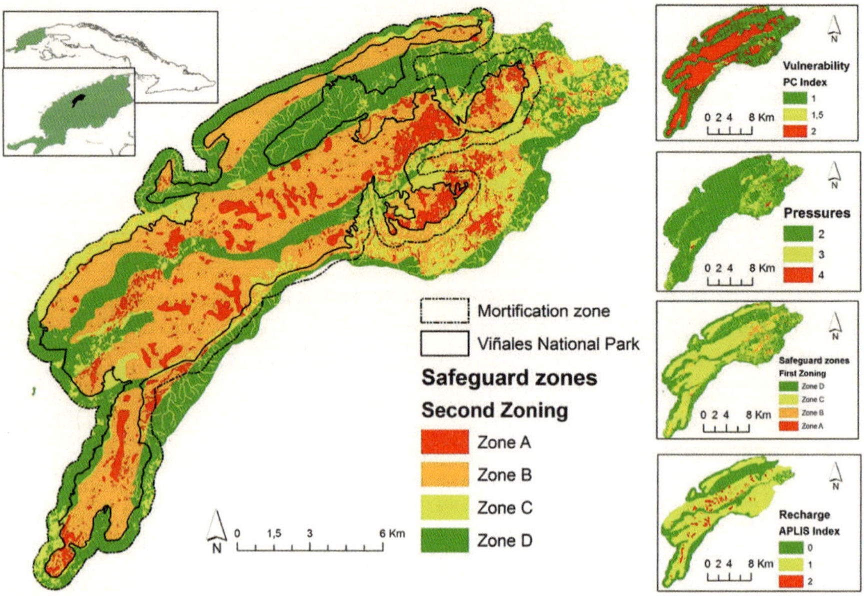

Fig. 12.2 Groundwater protection zones in Viñales National Park. The different zones respond to the criteria of Jimenez-Madrid et al. (2010)

12.5 Conclusions

Based on the methodology proposed by Jiménez-Madrid et al. (2010) a first approach to zoning groundwater protection of water resources in Viñales National Park is realized. The slight modification of the method allowed a better approach to zoning safeguard zones in the territory. Due to the karst nature, approximately 50 % of the territory requires strict and moderate protection. The binary behavior of the main karst system makes very complex the differentiation of the areas of the spring's basins, especially, due to the presence of annexed systems of recharge and diffuse drainage.

Nevertheless, it is necessary to extend the studies to potable water sources based on a best definition of the hydrogeological structure of the aquifer systems. For this assessment it is necessary take into account the karst network development and the subsequent determination of the inner and outlet zones. On other hand, for a better performance of a good response plan against contamination events, the energetic response of the aquifers, based on hydrograph and chemographs analysis, is necessary. These analyses, together with mathematical models tracer tests in different hydrological condition and geophysical methods can help in the validation of the methodology.

Acknowledgments The results were achieved due to the activities supported by the Management Plan of National Park Viñales, especially the investigation programs: "*Hydrogeology and vulnerability of karst aquifers in the National Park Viñales*" and "*Integrated studies of physical environment for implementation of treatment residual water systems in communities of National*

Park Viñales", this last of which was in coordination with the University of Alcalá and IMDEA Water Foundation.

References

Aldana C, Farfán H, Molerio L, Parase M (2007) Self-purification capability of the course or underground water in tropic humid karst mountain. Result of a tracer experiment in the Gran Caverna de Santo Tomás. Geophysical Research Abstracts, vol 9, 01841, EGU Assembly, Viena

Andreo B, Vías J, Duran JJ et al. (2008) Methodology for groundwater recharges assessment in carbonate aquifers: application to pilot sites in southern Spain. Hydrogeol J 16(5):911–926

Díaz C (1999) Geología y geomorfología del carso de la Sierra de San Vicente. Tesis de Maestría, Mención Geología Regional, Universidad de Pinar del Río

Díaz C, Farfán H (2007) Geomorfología e Hidrología de la Sierra de San Vicente, Viñales, Cuba. II Convención Cubana de Ciencias de la Tierra. Geociencias 2007. La Habana, Cuba. ISSN: 978-959-7139-83-6

Farfán H (2010) Evaluación de los recursos hídricos del Parque Nacional Viñales, Cuba. MSc Thesis, University of Alcalá

Farfán H, Molerio L, Díaz JE (2005) Aproximación al funcionamiento hidrológico del Valle de Viñales. Cong. 65 aniversario de la SEC. Libro de resúmenes y CD

Farfán H, Diaz C, Parise M, Aldana C (2010a) Scenarios of groundwater pollution in a karst watershed: a case study in the Pinar del Río province at Cuba. In: Andreo B, Carrasco F, Duran JJ, LaMoreaux JW (eds) Advances in research in karst media. Springer, Heidelberg/Berlin, pp 289–295

Farfán H, Corvea JL, de Bustamante I (2010b) Sensitivity analysis of A.P.L.I.S. method to compute spatial variability of karst aquifers recharges at the National Park of Viñales, Cuba. In: Andreo B, Carrasco F, Duran JJ, LaMoreaux JW (eds) Advances in research in karst media. Springer, Heidelberg/Berlin, pp 19–25

Farfán H, Corvea JL, de Bustamante I (2010c) Zonas de Protección de las aguas subterráneas de la cuenca del arroyo de Santo Tomás, Parque Nacional Viñales, Cuba. In: Goy JL, Cruz R, González A, Graña A, Cabero A (eds) Geomorfología y geología ambiental aplicadas a la gestión de espacios naturales protegidos. Dpto. Geología. Univ. Salamanca, España, pp 113–117

Farfán González H, Rodríguez C, Días Sosa JE (2009) Vulnerabilidad a la contaminación de las aguas subterráneas de la cuenca del Arroyo de Santo Tomás, Viñales, Cuba. MAPPING No. 137: 60–64. ISSN: 1.131-9.100

Gutiérrez R (1995) Condiciones geológicas del desarrollo diferenciado de la carsificación en la Cordillera de Guaniguanico, Cuba Occidental. In: Pulido-Bosh A, Fagundo JR, Rodrigues-Rubio JE (eds): El Karst y los acuíferos Karsticos. Ejemplos y métodos de estudio, Ediciones Osuna, Universidad de Granada, pp 11–26

Jiménez-Madrid A, Carrasco F, Martinez C (2010) The protection groundwater destined for human consumption in karst aquifers. Advances towards safeguard zones. In: Andreo B, Carrasco F, Duran JJ, LaMoreaux JW (eds) Advances in research in karst media. Springer Book, Heidelberg/Berlin, pp 299–304

Marín AI, Vías JM (2010) Practical session on groundwater recharge assessment in carbonate (karst) aquifers. In: Andreo B, Duran JJ (Coordinators): HIDROKARST –2010 – Groundwater recharge assessment and contamination vulnerability mapping in karst aquifers, p 51

Molerio LF (2004) El enlace absorción-descarga de la Gran Caverna de Santo Tomás: evidencias derivadas de un ensayo con trazadores artificiales. Ing Hidr y Ambiental XXV(3):22–26, La Habana

Molerio León LF (1995) Regionalización Hidrogeoquímica de las aguas subterráneas en la Sierra de Quemado, Pinar del Río, Cuba. Congr. Internac. LV Aniv. Soc. Espel. Cuba y Primera Reunión Iberoamericana. La Habana. Cuba, 94pp

Molerio León LF, Flores Valdes E (2003) Hidrogeología y Geomorfología carsica de Valle Ancon. Pinar del Rio, Cuba, p 14pp (Inédito)

Molerio LF, Menéndez A, Flores E et al. (1997) Potencial y recursos de aguas subterráneas en las zonas de montaña de Cuba. In: Arellano DM, Gómez-Martín MA, Antigüedad I (eds) Investigaciones hidrogeológicas en Cuba. LANKOPI, País Vasco, pp 211–225

Nguyet TM, Goldsheider N (2006) A simplified methodology for mapping groundwater vulnerability and contamination risk, and its first application in a tropical karst area, Vietnam. Hydrogeol J 14:1666–1675

Nuñez A, Panov V, Stelcl O (1965) Serie Espeleológica y Carsologica # 2. Academia de Ciencias de Cuba, Carsos de Cuba, p 47

Part IV
Design and Management Water Resources in Protected Areas

Chapter 13
Flood Hazard Mapping of the Yeltes River Basin for Planning of Natural Spaces "Los Batuecas-Sierra de Francia" and "Quilamas" and Their Area of Influence (Salamanca, Spain)

A.M. Martínez-Graña, J.L. Goy-Goy, and C. Zazo Cardeña

Abstract This paper presents a flood risk assessment of the Yeltes river basin (SO Salamanca, Spain), where the flooding and river overflow processes can lead to loss of life and property. The risk is determined by GIS and remote sensing techniques, based on the hazard map computed by a cartographic procedure comprising historical, hydraulic and hydro-geomorphological methods. The geomorphological method implements precise hydraulic modelling to establish the variations in the geometry of the canal in times of flooding and the contour conditions in general, defining geomorphologically active areas. The flood risk maps take into account the exposure and vulnerability, identifying areas with different degrees of risk and the need for protective measures (mainly structural). The existence of structural measures for prevention and mitigation have been taken into account, as they validate these risk maps, which themselves are non-structural measures and tools for effective planning and management of natural spaces.

Keywords Flood risk assessment • Hydraulic and hydro-geomorphological methods • Hydraulic modelling • GIS • Remote sensing • Salamanca • Spain

---

A.M. Martínez-Graña (✉) • J.L. Goy-Goy
Department of Geology, External Geodynamics Dept., Faculty of Sciences, University of Salamanca, Plaza de la Merced s/n, (37 008) Salamanca, Spain
e-mail: amgranna@usal.es; joselgoy@usal.es

C. Zazo Cardeña
Geology Section, National Museum of Natural Sciences, C/José Abascal n° 2, Madrid, Spain
e-mail: mcnzc65@mncn.csic.es

H. Farfán González et al. (eds.), *Management of Water Resources in Protected Areas*, Environmental Earth Sciences, DOI 10.1007/978-3-642-16330-2_13,

Fig. 13.1 Study area

13.1 Introduction

Water is a necessary and beneficial resource for society. Since primitive times, this has led people to settle in the alluvial plains, where optimum features for their needs were found: fertile soils, flat topography and water availability. As of this occupation, inundations or fluvial overflow processes led to loss of human lives and material goods, so the overflow risk is considered on the basis of the flood processes exceeding a certain threshold. An extraordinary runoff causes fluvial overflow when the flow exceeds the drainage channel capacity and water occupies the floodplain, which is a natural hydro-geomorphological phenomenon formed by the river to withstand or absorb flooding.

From the late nineteenth century, industrialisation gave rise to occupation of the floodplain and the resulting expansion of urban areas to lower river basin zones, leading to a social demand for structural or non-structural measures to protect against the risk of flooding. To this end, flood risk maps and cartography of isobaths and isotachs were drawn up for the Yeltes river basin within the Duero basin (Fig. 13.1), integrated in a comprehensive risk mapping in the area of influence of the natural spaces of Las Batuecas-Sierra de Francia and Quilamas. The implementation of protection measures (mainly structural) creates a “false sense of security” that intensifies land use, so the degree of development of a society grows in inverse proportion to its vulnerability. It is therefore necessary to establish an analytical methodology to help solve environmental and urban development flood risk problems, especially in natural spaces and adjacent areas.

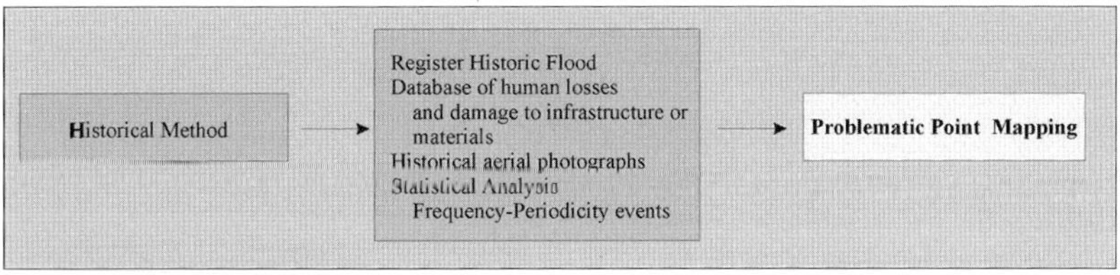

Fig. 13.2 Historical method procedure

13.2 Methodology

The methodology to be followed is based on establishing a mathematical model, in which a set of parameters (maximum rainfall, precipitation threshold, design flows according to return periods) required for flood risk modelling is obtained by calculation procedures. Basically, the combination of these parameters in the Yeltes river system is used to represent reality in a simplified form (Joly 1988). This procedure is used to assess the overall system performance, in this case fluvial, applying several hypothetical scenarios based on the common or fractal parameters that identify the system.

The flood risk analysis of the river Yeltes was performed estimating the risk of flooding from the hazard, exposure and vulnerability to flood or overflow processes in the margins of the river channel Yeltes. The hazard analysis is approached using three complementary and integrated methods: Historical Method, Historical-Hydraulic Method and Geomorphological Method Diez Herrero (2003).

The historical method (Fig. 13.2) is based on analysis of overflow and flooding events to establish hotspots that have been identified in different maps, or in databases, showing the floodwater heights reached, flow records, news in the media, photographs of events, etc. (IGME 2004). The regional problems of flooding over time are deduced from the study of historical floods in order to analyse the problems identified under current conditions and extrapolate the best solutions. The task thus consists of locating the zones most frequently hit by floods and collecting, classifying and organising the data collected to define the main causes that led to flooding, the most frequent types of damage and their relative magnitude.

This historical analysis cartographically locates floods of varying severity, both in time and space, including the affected area, flood date, duration, possible causes, damage attributable, solutions and remedies attempted after its occurrence and, of course, the sources from which data were obtained. In the case of the river Yeltes, the data are gathered from locations in the middle and lower reaches of the river where the runoffs from upstream tributaries accumulate, with low density population centres, and where the temporary incapacity of the main river to transport the flows brought by the floods leads to overflows. Proof of this is the presence of protective works in the form of defences (walls and ditches) that often collapsed during flooding, exacerbating the problem.

Finally, a map of historical floods was drafted for the Yeltes basin, based on historical records, showing the risk of flooding where the river Yeltes runs through

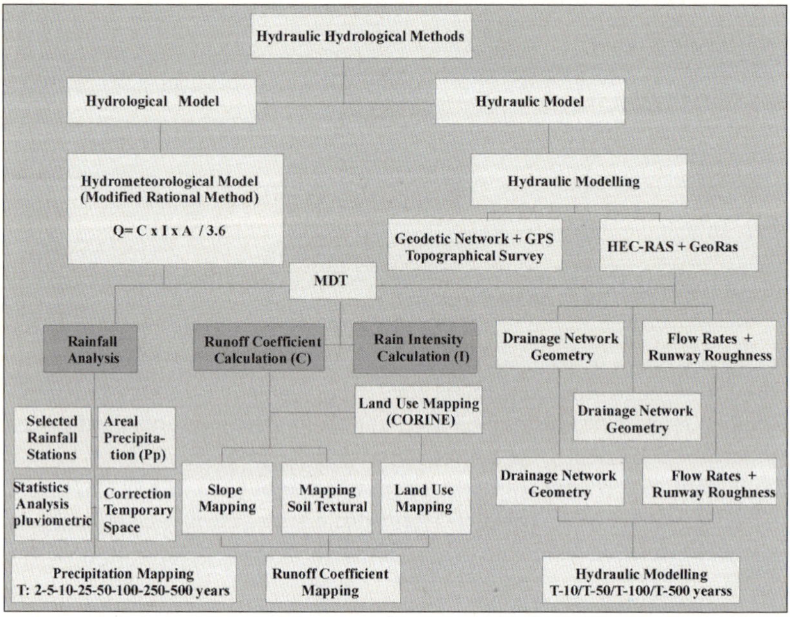

Fig. 13.3 Hydrologic-hydraulic method procedure

the town of Alba de Yeltes, Aldehuela of Yeltes, Puebla de Yeltes and its tributary Gavilanes as it passes through Sancti-Spiritus.

The Hydrologic-Hydraulic Method is a quantitative approach widely used in flood risk studies. The method is performed in two stages (Fig. 13.3): the first consists of hydrological modelling based on statistical flow analysis (graphic fit, parametric fit. . .), and hydrometeorological methods (rational method). In this initial stage, the overflows or peak flows carried by the river along its course are calculated.

In the second stage, hydraulic modelling is carried out, which can be set according to different levels of complexity: single-phase/two-phase, one-dimensional/two-dimensional, even/diverse, seasonal/variable. The aim of this second phase is to determine the water surface heights and flow speeds reached when circulating in a given channel.

To analyse the flood risk in the Yeltes river basin, it is necessary to know the maximum flow rates to subsequently be able to perform the hydraulic modelling with this contour condition. To determine these flows, given the absence of appraisals in the study area, hydrological modelling was done previously. Rainfall-runoff models were used in which, on the basis of analysis of the rainfall and physical characteristics of the basin, the maximum daily precipitation values are transformed and grouped by return periods into flood flows by means of the deterministic aggregation approach known as Modified Rational Method (Témez 1991). The flow calculation was done for each of the tributaries of the river Yeltes, considering these sub-basins independently. For this hydrological analysis with ArcGis 9.2, two extensions are used: "Arc HydroTools" and "3D Analyst".

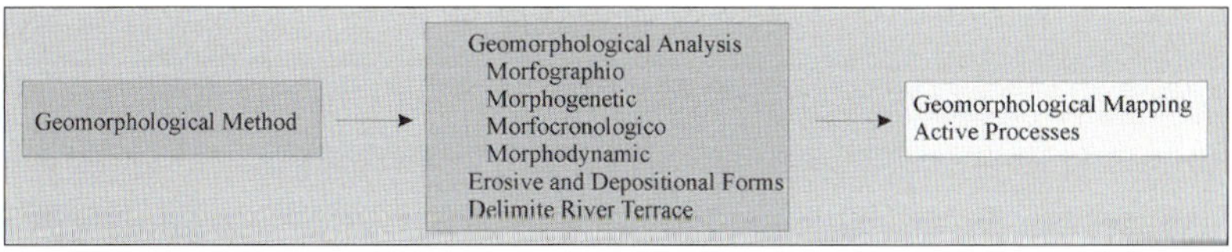

Fig. 13.4 Geomorphological method procedure

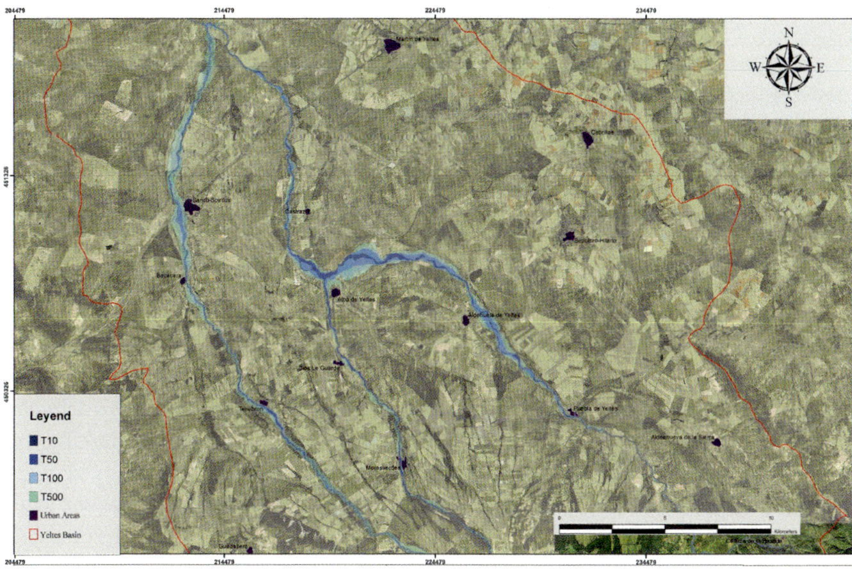

Fig. 13.5 Yeltes river basin flood mapping by different return periods (10, 50, 100 and 500 years)

The geomorphological method (Fig. 13.4) allows the wet perimeter of the different return periods to be specified, based on the analysis of different forms and deposits associated with the channel (Pinter et al. 2001), as well as any possible paleoinundation events that could be determined from the geological record or elements revealing the height of the water layer in the field after a flood, such as waste materials that "floated" on the water surface and became trapped by vegetation as the flow subsided.

13.3 Results and Conclusions

From the corrected precipitation values, obtained from raster rainfall maps according to tributaries and return periods, the parameters necessary to determine the rainfall intensity are calculated, such as: the hourly-daily rainfall intensity ratio and a series of basin-specific parameters for the calculation of concentration time. The intensity is then calculated for return periods of 10 years, 50 years, 100 years and 500 years. Finally, the different design flows are used to estimate the extent of the water surface for each return period considered, obtaining the risk mapping (Fig. 13.5).

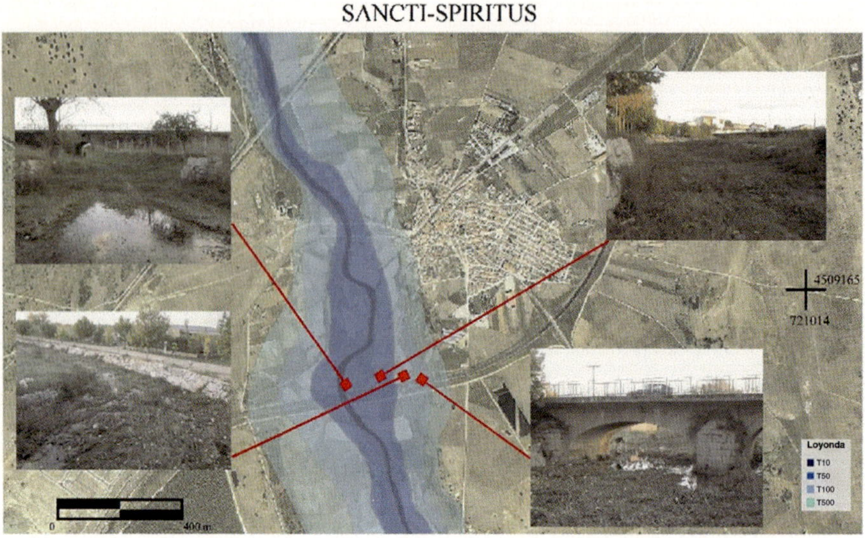

Fig. 13.6 Yeltes river basin flood mapping for different return periods (10, 50, 100 and 500) years

Gauging the disaster risks in different parts of Spain illustrates the need to delineate the flooded areas not only on the basis of hydrological criteria but also taking into account geological and geomorphological criteria, using maps of Quaternary geology, geomorphology and active processes, although due to questions of scale the latter cannot provide the zoning of the floodplain required for certain projects and/or regional planning. Nevertheless, they do enable definition of the main channel or river bed, analysing its morphology and dynamics as well as complementary information on other flood-associated phenomena: landslides, erosion, piping and human actions on the drainage channels (jetties, dams, quarries, landfills and dumps, dams, etc.). The legal standard applied in Spain to define the channel is a hydrologic approach shown to be unsuitable to determine flood areas, as other events and field observations have to be considered, in addition to the theoretical flow of normal maximum flooding, using fluviomorphological, environmental and social criteria. With the modifications generated by this method, the flood risk areas are more accurately defined.

The Yeltes river basin, based on historical flood maps obtained, runs significant risk of flooding, presenting zones where the urban areas, infrastructure and property are highly exposed and vulnerable to the different return periods. Analysing the results of the flood risk mapping in different localities and superimposing them on the 1:2,000 orthophoto, the vulnerability and exposure of different elements (Fig. 13.6) can be observed. This is useful for the management of natural areas to establish the protection measures provided for the exposed and vulnerable zones and determine the anthropogenic expansion and level of protection.

In the Yeltes river basin, the most common structural measures are correction and regulation of riverbeds by cleaning and dredging of channels, increasing the transport capacity of the river during floods, as it increases the longitudinal slope and section of the channel, causing an increase in the flow drained, being a measure recommended in the headwaters of the Yeltes, especially after a flood. The protection of river channels is also important, analysing the drainage capacity of the existing crossing works and examining the additional protection works that may be required on the basis of the calculations developed in the hydrometeorological and hydraulic modelling corrected with the geomorphological method. These structural measures would be suitable for return periods corresponding to regular and moderately severe flooding, and the non-structural measures against extraordinary floods, increasing river channelling for return periods of 100 years to 500 or reducing to 50 according to their vulnerability, and, finally, in agricultural areas a protection level of 50 years or less, depending on the value of the crops.

Acknowledgments Projects CGL 2008-03998/BTE, SA041A08 (JCyL) and CGL2008-04000/BTE.

References

Diez Herrero A (2003) Geomorfología e Hidrología fluvial del río Alberche. Modelos y SIG para la gestión de riberas. Serie Tesis Doctorales nº 2. Edita Instituto Geológico y Minero de España (Ministerio de Ciencia y Tecnología), Madrid

IGME (2004) Pérdidas por terremotos e inundaciones en España durante el periodo 1987–2001 y su estimación para los próximos 30 años (2004–2033). Ed. Consorcio de compensación de seguros, Madrid

Joly F (1988) La Cartografía. Ed. Oikos-Tau, Barcelona

Pinter N, Thomas R, Wollsinski JH (2001) Assessing flood hazard on dynamic rivers. EOS Trans Am Geophys Union 82:333–339

Témez JR (1991) Generalización y mejora del método racional. Dirección General de carreteras de España, Madrid

Chapter 14
Estimation of Water Excess in the Medium and High Basins of Yanuncay River (Cuenca-Ecuador) by Approximation of Monthly: Distributed Water Balance

J. Fernández de Córdova W. and J.A. Pascual

Abstract In this study, the water excess generated during a 10-year period (1998–2008) in the high and medium basins of Yanuncay River is approximately determined. The distributed water balance method was used, which analyzes the interaction between all the different parameters that form part of the water cycle in nature. To create the model, basic parameters such as precipitation, temperature, soil type and superficial covering are required. Furthermore, by using computer software such as Microsoft Office Excel and Geographic Information Systems (GIS) it is possible to obtain monthly data showing the water excess and to generate thematic maps, which allow for an effective monitoring of the behavior of the medium and high basins of Yanuncay River.

Keywords Water balance method • Water cycle • Modeling • GIS • Ecuador

14.1 Introduction

The water balance method is one of the most used procedures in hydrology to solve the different scenarios related to the movement of water in nature. This method permits the evaluation and quantification of the different elements that are part of the hydrological cycle in a specific period of time.

A hydrological balance is an equation representing the equilibrium existing between contributors to a determined basin, the volume of water leaving the watershed, the volume retained in the soil and the water consumed by vegetation.

J. Fernández de Córdova W. (✉)
Empresa Pública Municipal de Telecomunicaciones, Agua Potable, Alcantarillado y Saneamiento de Cuenca – ETAPA EP, Avenida Octavio Chacón 1-55, Cuenca, Ecuador
e-mail: jfernand@etapa.net.ec; javierfdezdec@hotmail.com

J.A. Pascual
Centro de Investigaciones sobre Desertificación (CIDE), Camí de la Marjal, s/n 46470, Albal, Valencia, España
e-mail: juan.a.pascual@uv.es

H. Farfán González et al. (eds.), *Management of Water Resources in Protected Areas*, Environmental Earth Sciences, DOI 10.1007/978-3-642-16330-2_14,

Thus, this equilibrium can be determined by quantifying the amounts of water entering (precipitation) the system, the amounts leaving (evaporation, transpiration and water excess) and the variation of the amount being retained by the system.

Several authors have studied the different factors of the water balance method; however, the method has been enforced due to the creation of a new concept known as Potential Evapotranspiration, introduced by Thornthwaite in 1948–1955. Thornthwaite found a formula to calculate the monthly value of potential evapotranspiration from climatologic data such as temperature and time in hours of sun exposure. This provided a procedure that allows the calculation of the hydrological balances on a monthly basis in a determined place and in a determined period of time, by knowing the system input parameters (rainfall), evaporation losses, vegetation transpiration and the amount of water retained by the edafic profile of the basin.

To be knowledgeable of the structure of water balance is fundamental for an efficient handling of the hydrological resources of a basin. It provides essential information leading to a better use of water stored in dams and forecasting consequences that could be the result of artificial modifications to the course of the river, lakes and underground basins. It also permits the indirect evaluation of any unknown components in the equation by quantifying the difference with the known parameters (Instituto de Hidrologia de Espana Unesco 1981).

The water balance method can be used to indirectly measure evapotranspiration, moisture content of the soil and the water excess generated in a Basin (Pascual 2002). Furthermore, it allows the introduction of temporary and spatial variables within the balance, such as land use and the different types of superficial covering that can be found in a basin. It is then possible to identify the zones generating the largest amounts of water excess and the zones with great capacity of water retention; thus, existing protected areas can be ratified and new protected areas may be established.

14.2 Area of Study

Yanuncay River is one of the main tributaries of Paute River. Its resources are used for drinking water, irrigation, recreational purposes and in the near future for the generation of electricity. Therefore, it is important to study the behavior of its watersheds.

This study analyzes the medium and high zones of the Yanuncay River basins, which constitute the main source of supply of one of the drinking water treatment plants that ETAPA EP (Empresa Publica Municipal de Telecomunicaciones, Agua Potable, Alcantarillado y Saneamiento de Cuenca) owns (Fig. 14.1).

The area of study is located approximately 15 km west of Cuenca in the towns Banos and San Joaquin. Cuenca is the capital of Azuay, one of the Provinces in which Ecuador is subdivided. Cuenca's total population (includes urban and rural) is approximately 400–17,000 people, which is the third largest population in the country.

This zone has a total area of 330 km^2, representing nearly 80 % of the total area of Yanuncay's watershed (418 km^2). It has a perimeter of 96 km and it is situated between 2,960 and 4,340 m height above sea level.

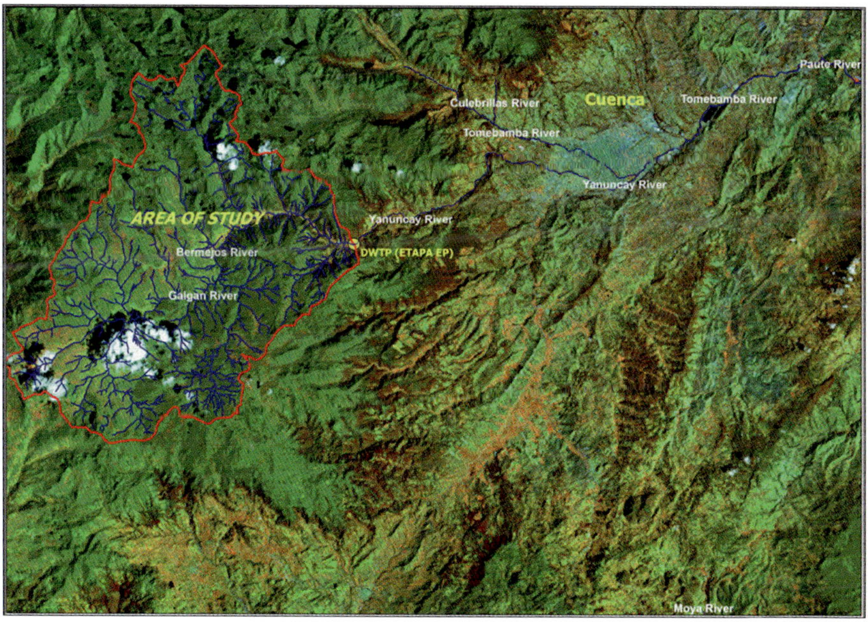

Fig. 14.1 Location of the area of study

Most of the ecosystems located in this area are of the moor type, which can only be found when over 3,500 m above sea level in Ecuador. The vegetation that can be predominantly found in this type of ecosystems is pads, mosses, aguarongos and grassland. These kinds of environments retain large amounts of water, which contributes to the formation of lakes, marshes and wetlands, similar to the ones found in the medium and high basins of Yanuncay River.

14.3 Methodology

The water balance method is based upon an equation that relates, in a simple way, to the amounts of water of different components of the hydrological cycle and it is expressed as follows:

$$\text{Inputs} - \text{Outputs} = \text{Storage Variation} \tag{14.1}$$

Now if it is considered a determined interval in a closed basin, i.e., where there is no water interchange with another basins, then the previous equation can be expressed as shown in Fig. 14.2.

In Fig. 14.2 it can be clearly observed that if precipitation, evapotranspiration and the variation of water storage are known then the water excess can be obtained in the system.

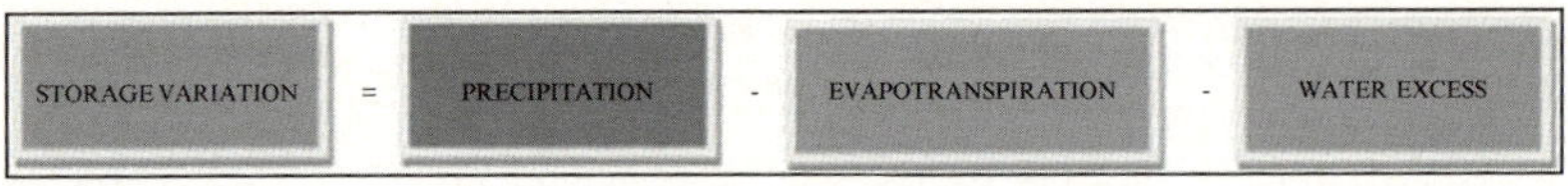

Fig. 14.2 Watershed hydrological balance (Martinez et al. 2006)

To estimate the water excess in this study, a Distributed Parameter Model was implemented which was based on a Geographic Information System (GIS). The main advantage of using this system is the spatial consideration given to the different variables intervening on the hydrological process of a watershed.

To apply the method of water balance, it is necessary first to have monthly data showing the climatic characteristics of the area of study, i.e., precipitation and temperature. Furthermore, it is necessary to know what types of soils and superficial coverings are present in the area, which integrates as to obtaining of the maximum potential capacity of water retention.

Basic information of this study:

1. Precipitation: With the use of Paute's River unified hydrometeorology network, which is maintained and operated by ETAPA EP, it was possible to obtain tabular data of monthly rainfall from August 1997 to October 2009 from two stations located within the zone of study (Cancan and Yanuncay in Pucan), followed by the generation of a map showing the distribution of precipitation in the area (Fig. 14.3).
2. Temperature: Yanuncay's watershed is not provided with the previously mentioned type of information; therefore, the tabulated data was taken from the stations of El Labrado and Marianza (Cajas) that are located in nearby watersheds, Machangara and Tomebamba rivers, respectively. Based on this data potential evapotranspiration was obtained by using Thornthwaite's formula. It was considered that ETP does not vary throughout the area of study.
3. Soil type: Shapefile data in 1:50,000 scales were facilitated by the Department of Management of Soil and Water of the University of Cuenca (PROMAS). Three main types of soils and one rock bed were identified (Fig. 14.3). To every type of soil the values of field capacity (CC) and permanent wilting point (PMP) were assigned.
4. Superficial Covering: Based on the 2001 map obtained from the Geographic Information System (GIS) for the analysis of spatial information of the Paute's river watershed, prepared by the University of Azuay (Fig. 14.4) (http://www.uazuay.edu.ec/promsa/paute/home.htm), eight different types of superficial coverings were identified. These data are required to determine the maximum potential capacity of moisture retention of the relation soils/superficial-covering (Cmrh).

The maximum potential capacity of moisture retention of the relation soils/superficial-covering is the result of the product between the depth of the roots (mm) and the difference (percentage) between the field capacity and the permanent wilting point (Pascual 2002). This product is calculated for all of the polygons that can be obtained from the combination of maps shown on Figs. 14.3, 14.4 and 14.5. The latter allows the determination of the spatial distribution of precipitation, the different types of soil and the superficial covering in the area of study.

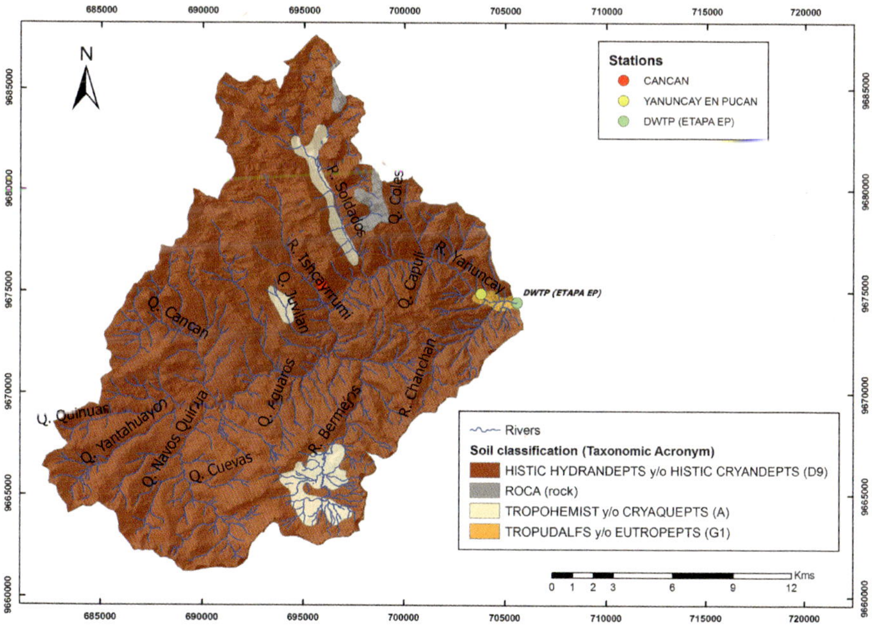

Fig. 14.3 Distribution of precipitation map in the zone of study (*top*). Types of soils in the area of study (*bottom*)

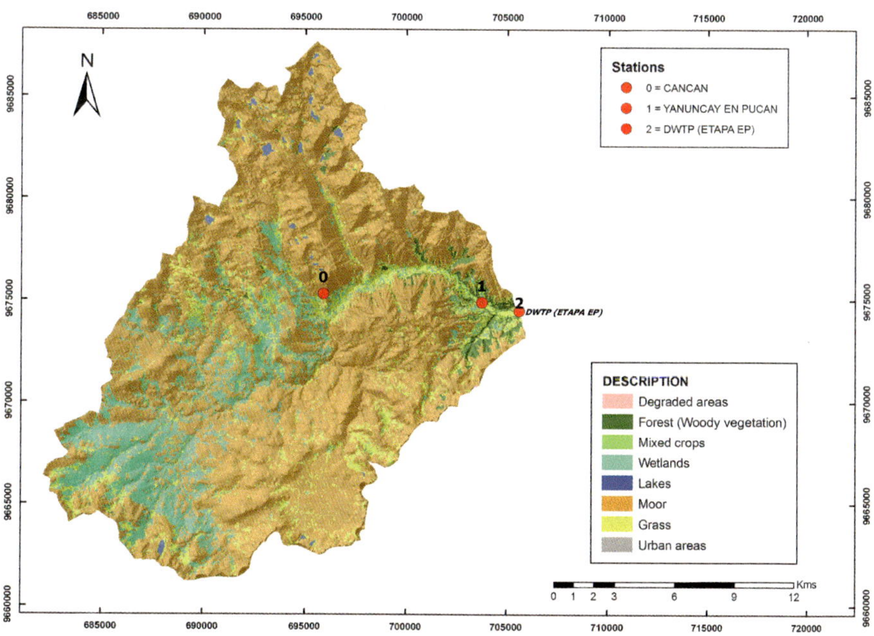

Fig. 14.4 Superficial covering in the area of study

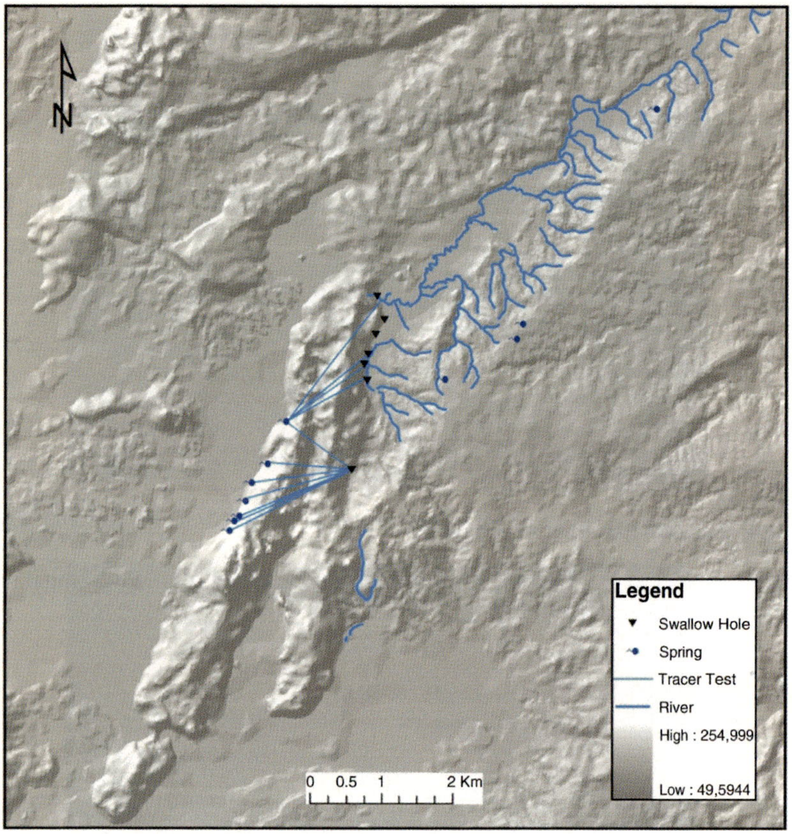

Fig. 16.5 Hydraulic connections defined by tracer tests

connect the entering streams; and huge lakes could last several days upstream, changing the flow patterns and the nutrient loads to the system. The Santo Tomás Cave System is a well-integrated 47 km of connected underground galleries; however, tracer tests (Molerio et al. 2004a) have shown that this figure could be duplicated. Figure 16.5 shows the interconnections discovered by means of tracer tests and shows the hydraulic relations among the components of the cave system.

16.4.3 The Physical and Biogeochemical Hydrodynamics

The most important features are: (1) the emission and assimilation of nutrient and contaminant non-conservative loads that commonly come from the upstream rural or peri-urban communities discharging poor or non-treated wastewaters; (2) the variable self-purification capability of the active cave system; and (3) the variation

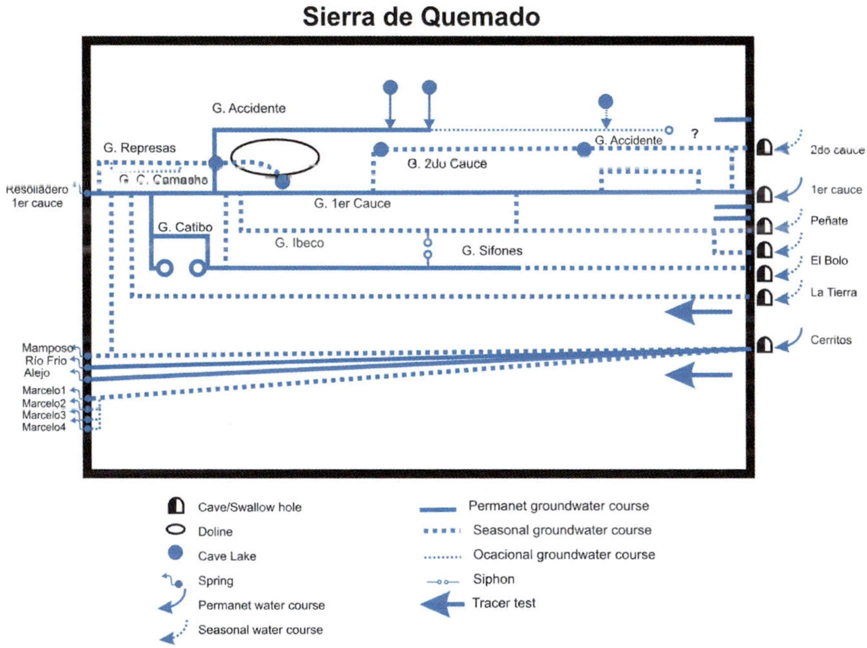

Fig. 16.6 General scheme of the explored and mapped underground connections at the Sierra de Quemado (Farfán et al. 2009)

and variability of the chemical composition of waters due to the residence time in the aquifer, the mixing length and the contact time rock-water.

16.4.4 The Local Level of Social and Economic Development

The local society is developed as a rural low-income community concentrated upstream of the absorbent slope and dispersed downstream of the discharge slope. No population lives in the Sierra de Quemado strictly. Water demand is insufficiently satisfied and domestic wastewater has practically no treatment (Fig. 16.6).

16.4.5 The Social and Natural Vulnerability and Hydrologic Risks

Social vulnerability and natural risks are more evident in the emissive (discharge) slope because of the dispersion of the population, the occupational framework and the types of housing (Molerio 2004b). Water demand is satisfied individually by familiar collection, and storage is not supervised by health activities. This led to generally insufficient water supply and to a low drinking water quality. That is not the case of the

Community of El Moncada, which is a system of municipal water supply although not very efficient operated. Isabel Maria is the most vulnerable community to water quality deterioration, but not to flood events. In this case El Moncada is more vulnerable because under certain conditions the cave system does not have enough capability to evacuate the floodwaters, and backflow could reach the community (as it has occurred during some hurricanes).

16.5 Conclusions

The effectiveness of any instrument of environmental management has to be founded in the adequate knowledge of the system where it will be applied. The following knowledge instruments should be taken into account:

- The laws governing the organization of the surface runoff.
- The local development of the hydrological karst system
- The physical and biogeochemical hydrodynamics.
- The local level of social and economic development
- The social and natural vulnerability and hydrologic risks.

The main vulnerability to contamination is located at the emissive slope and the most important stress comes from horizontal flows. No human stress is identified in the top of the hills, but in the slopes.

Well-integrated cave (flow) systems contribute to the improvement of input waters of low quality in terms of Dissolved Oxygen. On the other hand, low meandering and/or integrated systems do not produce any improvement in water quality and constitute a strong hazard to the communities living at the discharge slope.

References

Farfán González H, Días Guanche C, Parise M, Corvea Porras JL (2009) Inventario y caracterización de los escenarios de peligro a la contaminación de las aguas de la cuenca del Arroyo de Santo Tomás, Viñales, Cuba. MAPPING No. 135: 66–70

Furrazola-Bermudez G, Sánchez Arango R, García R, Basov VA (1978) Nuevo esquema de correlación estratigráfica de las principales formaciones geológicas de Cuba. "La Minería en Cuba" 4(3): 30–35

Molerio León LF (2004a) El enlace absorción-descarga de la Gran Caverna de Santo Tomas: evidencias derivadas de un ensayo con trazadores artificiales. Ing. Hidr. y Ambiental, La Habana, XXV(3): 22–26

Molerio León LF (2004b) Indicadores de vulnerabilidad de acuíferos cársicos. Ing. Hidr. y Ambiental, La Habana, XXV(3): 56–61

Piotrowska K (1987) Las estructuras de nappes en la Sierra de los Organos. In: Pszczlhowski A et al (eds) Contribución a la geología de la provincia de Pinar del Río. Edit. Cien.-Téc, La Habana, pp 85–156

Psczolkowski A (1987) Secuencias miogeosinclinales de la Cordillera de Guaniguanico. In: Pszczlhowski A et al (eds) Contribución a la geología de la provincia de Pinar del Río. Edit. Cien.-Téc, La Habana, pp 4–84

Rodriguez-Becerra M, Espinoza G, Wilks D (eds) (2002) Gestión ambiental en América Latina y el Caribe: evolución, tendencias y principales prácticas. Banco Inter. Desa., Washington, p 286

Chapter 17
The Relationship Between Surface Waters and Groundwaters in the Coastal Wetlands of Campo de Dalías (Almería, SE Spain) and Their Importance for Sustainable Water Management

F. Sánchez-Martos and L. Molina Sánchez

Abstract Water resources management in coastal wetlands requires knowledge of the degree of interdependence between groundwater and terrestrial ecosystems. This is especially so in semi-arid areas where surface inflows are restricted, marine influence is marked and the evaporation rate is high. In the Cerrillos-Punta Entinas wetland, the surface water quality is highly variable. Using hydrogeochemical tools, the main processes that favor the diversity of water types are described, arising from: the presence of salt deposits on the pool beds, the predominantly-marine origin of the water, and the local influence of groundwater. Understanding the influence of each process assists in determining how best to manage the water resources of this protected natural space in a sustainable way.

Keywords Water resources management • Coastal wetlands • Groundwater • Surface water quality • SE Spain

17.1 Introduction

There are several coastal wetlands in the Mediterranean basin whose size and development are not solely a result of their marine influence. Groundwater-surface water interrelationships are particularly important to understand these cases, especially in semi-arid areas where surface water inflows are very restricted. Given their coastal situation, the marine influence on water chemistry is marked, which makes the analysis of possible relationships with groundwater very difficult. In addition,

F. Sánchez-Martos (✉) • L. Molina Sánchez
Department of Hydrogeology, University of Almería, Campus Universitario, Almería 04120, Spain
e-mail: fmartos@ual.es

H. Farfán González et al. (eds.), *Management of Water Resources in Protected Areas*, Environmental Earth Sciences, DOI 10.1007/978-3-642-16330-2_17,

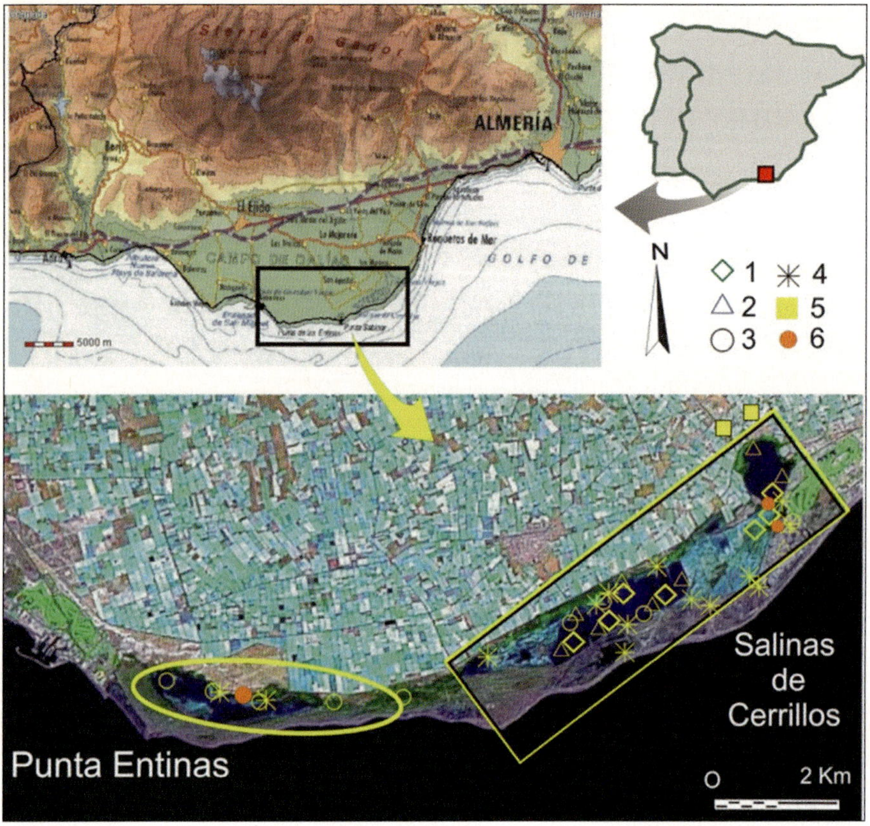

Fig. 17.1 Location of coastal wetlands in the Campo de Dalías (Punta Entinas and Cerrillos salt flats) and of the sampling points for the three data series used. Surface waters (1: 1985, 2: 1996, 3: 2005, 4: 2009). Groundwater (5). Mean water level in the ponds (6)

the active littoral dynamics need to be taken into account, as well as the intense anthropic activity, as in many cases the coastline has been modified by salt flats exploitation.

The Water Framework Directive (WFD) requires terrestrial aquatic ecosystems that are directly related with groundwaters to be defined. Even the quality objectives of the water bodies are linked to sustaining the environmental demands of these terrestrial ecosystems. These two requirements mean that the degree of dependence of wetlands on groundwater resources needs to be understood.

Given these requirements, the objective in the present study is to investigate the degree of interdependence between the surface waters and the groundwater flow to the Campo de Dalías coastal wetlands. A characteristic of these wetlands is their highly variable salinity (Fig. 17.1).

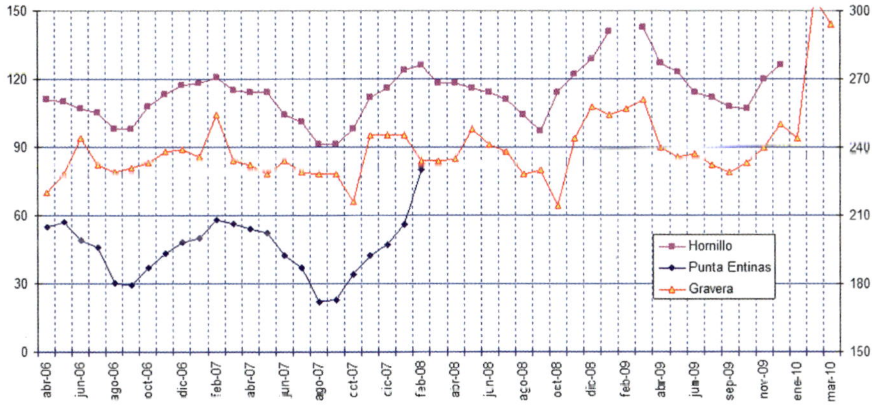

Fig. 17.2 Evolution of water level at three monitoring points in the wetlands. The situation of these points is given in Fig 17.1 (Data: Andalusia Wetlands Plan (Consejeria de Medio Ambiente, Junta de Andalucia))

17.2 Materials and Methods

The study area comprises two coastal wetlands ("Punta Entinas" and "Salinas de Cerrillos") situated on the coastal margin of the Campo of Dalías. Together, they constitute the "Paraje Natural Punta Entinas – Sabinar", a natural space (nature reserve) protected under regional legislation (Junta of Andalusia, Fig. 17.1). The wetlands consist of two partially-silted tidal lagoons, separated from the sea by sandbanks and dunes. Flooding of the Cerrillos Salinas (salt flats) is highly variable; there are two main lagoons of permanent water, while the remainder of the area contains small seasonal ponds that depend on rainfall. From 1948 until the end of its active life, the salt works were used as pre-concentration lagoons. Normally, the concentration process here fell short of precipitating gypsum, although in dry periods the complete precipitation sequence was achieved (Dronkert 1985). When salt extraction ceased at the end of the 1980s, seawater was no longer pumped into the salt flats and more than 70 % of the flooded area was left dry. Today, only the lagoons at the western end of the Salinas de Cerrillos are continuously wet throughout the year and unaffected by anthropic action. Water level data (Fig. 17.2) for three sampling points show seasonal cycles, with the lowest levels in summer and the highest, in spring. Of the three sites, the amplitude of variation at the point in Punta Entinas is wider than at the other two monitoring stations located in the extreme east of the wetlands.

Lower and Middle Pliocene deposits outcrop in the vicinity of the wetlands that consist of thick marly series covered by regressive deposits (sandy marls, sands and calcarenites). Continental Pleistocene deposits correspond to the distal facies of a series of alluvial fans and to very fine, red silts containing a limited detritic fraction. Marine Quaternary deposits comprise four Tyrrhenian beds that extend in a band parallel to the coast, 10–20 m above sea level. These determine, for the most part, the current morphology of the area (Goy and Zazo 1980). Other Quaternary deposits occur along the coast that are linked to recent coastline dynamics: salt pans, salt marshes and lagoons correspond to silty and muddy deposits occupying the low-lying zones; while sand dunes, wholly or partially fixed by vegetation, cover former marine terraces that have been exploited for agriculture; accumulation beaches extend parallel to the narrow fringe of the current beach. Together, these landforms make up an area of very recent, dynamic geology (Goy et al. 2003). To the north of this protected Natural Space are outcrops of Pliocene calcarenites. These are covered by marine terraces that form the closest aquifer formations to the wetlands. Throughout this sector are outcrops of Pliocene marls, which comprise the impermeable base of the aquifer. Surface runoff is very limited because of the small catchment, shallow gradient and the intense impact of greenhouses that surround this Natural Enclave.

17.3 Data

Three datasets were used in this study. The first survey was undertaken in 1985 (Dronkert 1985) and included 15 sampling points at the eastern end (Salinas de Cerrillos) during the final stages of salt extraction. The second study (Molina and Sánchez Martos 1996) looked at annual changes at ten sampling points around Cerrillos. The latest dataset comprised two sampling surveys (September 2005 and May 2009) and encompassed both Punta Entinas and the Cerillos salt flats, in order to characterize the nature reserve as a whole. To characterize the groundwater beneath the wetlands, five aquifer sampling points in the Balerma-Las Marinas aquifer were considered. Data was collected for the following parameters: electrical conductivity, Cl^-, SO_4^{2-}, Br^-, HCO_3^-, NO_3^-, Na^+, Mg^{2+} and Ca^{2+}. A summary of these parameters is given in Table 17.1 and Fig. 17.2.

17.4 Results and Discussion

The box-plot diagrams in Fig. 17.3 indicate the variability of ion content in the wetlands. These waters have a sodium chloride facies – due to the marine environment in which they have formed, though many of the ions are present in concentrations exceeding that of seawater due to the concentration of salts by evaporation. Even so, some data show values less than seawater. The groundwater is also sodium chloride type, but has a lower ion content (Table 17.1). The hydrochemical variability is highlighted and all surface water samples have a sodium chloride facies. The groundwater, with its lower saline content, is closer to a mixed facies and clearly different from the surface waters (Sánchez Martos et al. 2007).

Ion ratios were calculated to interpret the evolution of a series of surface waters, considered as mixtures of waters at the two extremes of salinity. The relative distribution of SO_4^{2-} and Cl^- (Figs. 17.4 and 17.5A) shows the general model of salinity: the most saline waters have a higher ion content than salt water and are very disperse. Their mineralization is greatly influenced by the concentration of salts that occurs due to evaporation of seawater and to the presence of salts in the beds of the ponds. This is the case for the waters sampled in 1996 in the Hornillo

Table 17.1 Statistical summary of physico-chemical data for surface water and groundwater (GW: groundwater, SW: surface water) for the three data series (ion content: mg·L^{-1}; EC: electrical conductivity in μS·cm^{-1})

		E.C.	Cl^-	SO_4^{2-}	HCO_3^-	Na^+	Mg^{2+}	Ca^{2+}	Br^-
SW 55 data	MAX	194,000	120,214	10,141	945.6	40,200	4,693	2,164	968
	MIN	8,320	2,185	514	111	1,437	136	100	5
	MED	77,942.7	35,986.6	4,298.9	243.9	17,941.2	1,948.1	815.1	142.9
	DEST	51,750.7	27,300.3	2,487.7	152.5	11,196.5	1,410.3	524.2	226.2
GW 10 data	MAX	5,660	1,610	382	392	1,157	158	175	4.3
	MIN	2,580	549.5	158.5	197.4	312.9	55.9	48.1	2.5
	MED	3,725.0	948.4	262.3	301.5	559.4	108.0	103.8	3.5
	DEST	1,135.5	359.9	75.0	60.8	274.8	32.8	35.5	0.9

MAX maximum, *MIN* minimum, *MED*: mean *DEST*: standard deviation

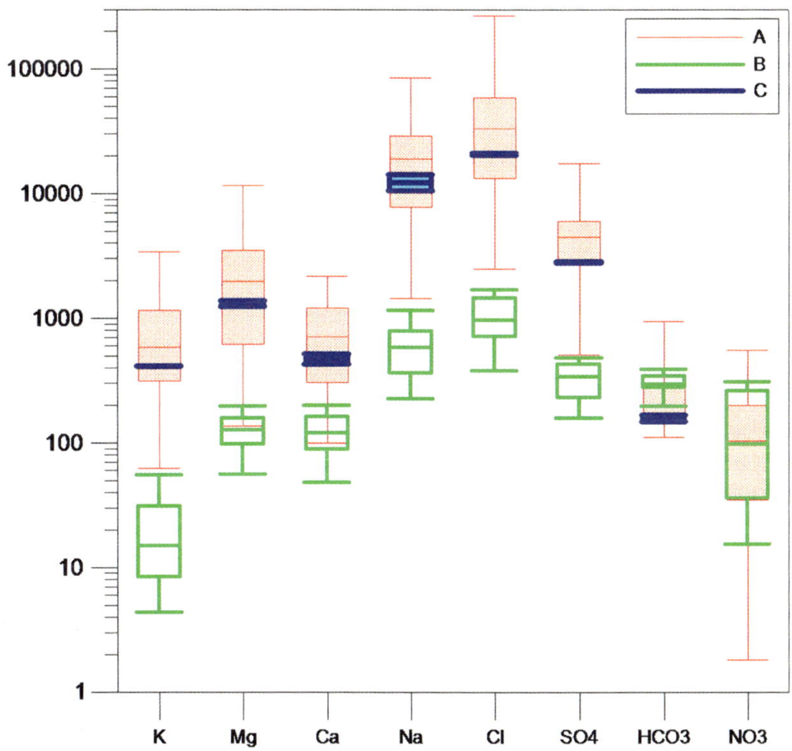

Fig. 17.3 Box-plot diagram for individual data sets: (A): surface waters; (B): groundwaters and (C): seawater. The range, 25 % and 75 % percentiles and mean values are indicated

lagoon, situated at the eastern end of the Cerrillos salt flats, which gave the highest SO_4^{2-}/Cl^- ratio of all. The SO_4^{2-} concentration was double that of seawater, indicating that there must be an additional source of SO_4^{2-}. This lagoon was used as the evaporator pond, and salinity was sufficiently high for gypsum to precipitate out. Of the remaining sampling points where salinity exceeds that of

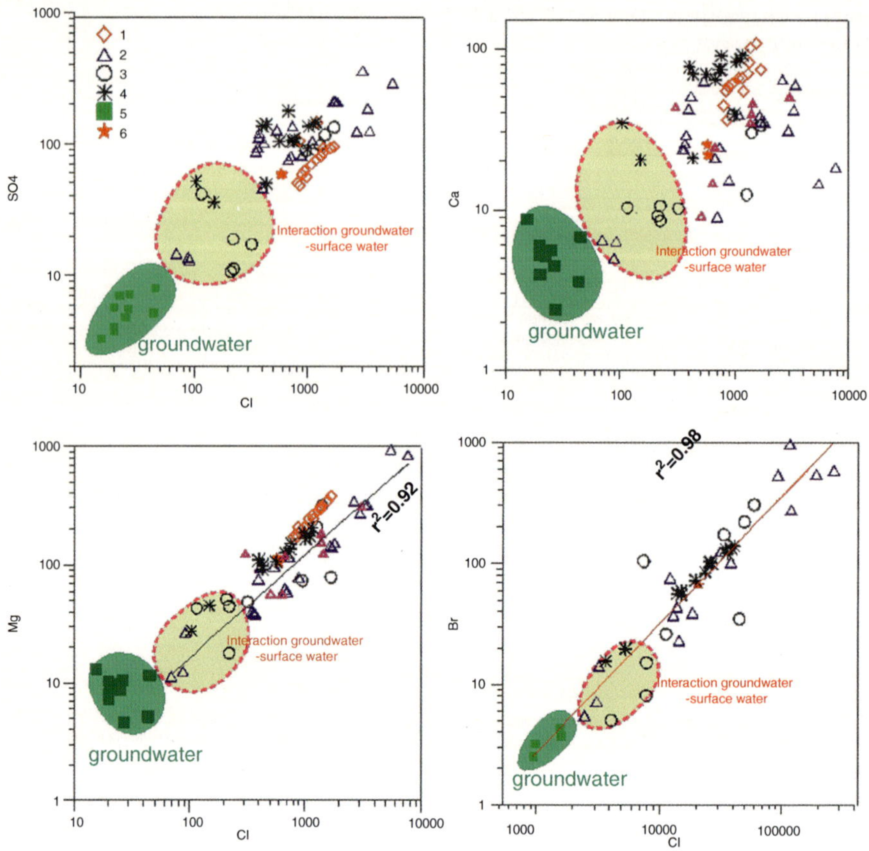

Fig. 17.4 Relationship between SO_4^{2-} and Cl^- (A); Ca^{2+} and Cl^- (B); Mg^{2+} and Cl^- (C), and Br^- and Cl^- (D). Symbols: Surface waters (1:1985; 2: 1996; 3:2005, 4:2009) 5: groundwater, 6: seawater. Data in (meq·L^{-1} and Br^- in mg·L^{-1})

seawater, there is a wider dispersion and this is related to the dissolution of chloride salts in the beds of other ponds. In summer, evaporation is complete and precipitation of highly soluble Cl^- and Mg^{2+} occurs as $MgCl_2$ and $MgSO_4$, for example (Dronkert 1985). When the salts re-dissolve during wet periods, it is mainly Mg^{2+} and Cl^- that are released to the brine, leaving the sulphate as a precipitate. For this reason, in the most saline samples, the Mg^{2+} and Cl^- (Fig 17.5C) has a high coefficient of correlation ($r^2 = 0.92$).

The distinct water chemistry is demonstrated by Br^- and Cl^-, which are concentrated from seawater to different degrees (Richter and Kreitler 1993) as a result of their distinct interactions with the rock matrix (Davis et al. 1998). Figure 17.5D indicates the close relationship between Cl^- and Br^- ($r^2 = 0.98$) that is due to the marine origin of the water. Points whose salinity is less than seawater are plotted in intermediate positions between seawater and the groundwater from the aquifer that lies south of the wetlands, as indicated by Cl^- and SO_4^{2-}.

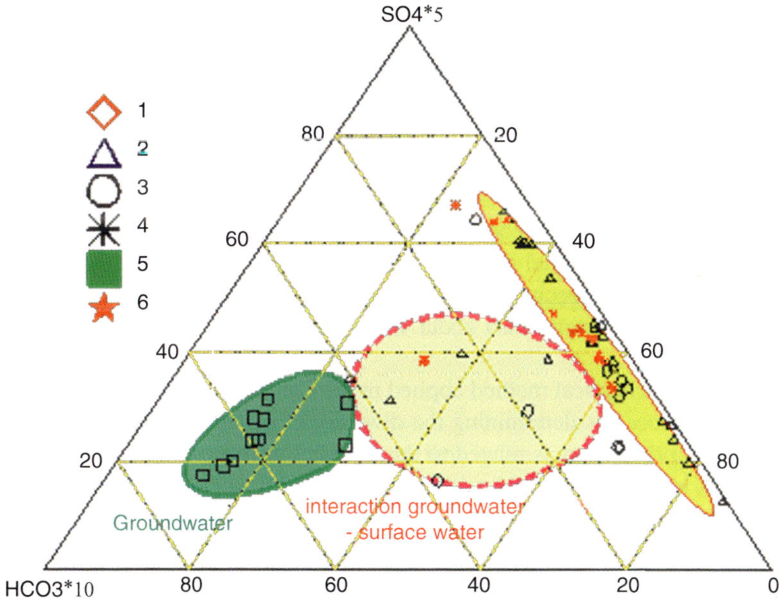

Fig. 17.5 Triangle diagram. This diagram exaggerates the values of HCO_3^{-2} (×10) and SO_4^{2-} (×5) Identification of data: surface waters (1:1985; 2: 1996; 3:2005, 4:2009) 5: groundwater 6: seawater

The triangle diagram can be used to better discriminate points that may be influenced by groundwater flows: the diagram exaggerates the axes, so giving a wider dispersion of points than in a classic triangle diagram. In Fig. 17.5, values of HCO_3^{-2} and SO_4^{2-} are exaggerated in this way (by ×10 and ×5, respectively) to separate the groundwater data towards the HCO_3^{-2} vertex because of their greater carbonate content. Another group of surface water samples is also highlighted, which lie in an intermediate position between groundwater and seawater – these represent the two extremes of water inflow to the wetlands and reflect the varying influence of groundwater and surface waters.

That the salinity of the wetlands is so variable, is strongly related to the marine salts present in the beds of each of the ponds. In addition, the lower salinity recorded in the eastern sector of the Salinas de Cerrillos (Hornillo Lagoon) and in some parts of the western reserve (Punta Entinas) demonstrates the influence of groundwater mixing.

17.5 Conclusions

The water chemistry of these coastal wetlands shows a marked variability due to a diverse series of factors. They are areas with very active geological dynamics, and this has favored the development of endorrheic basins where surface water accumulates,

18.1 Introduction

18.1.1 Efforts Pro-wetlands

Curiously, regardless the actual state of wetlands, there is a great deal of concern at the international level about their values, benefits and the need to conserve them. It is the only ecosystem that has an international convention (Ramsar) signed by 159 countries agreeing to protect and maintain wetlands given their relevance. Scientists are even imitating their functions in artificial wetlands as an alternative for water treatment among other valuable uses.

Moreover, wetlands are also part for the implementation and success of strategic programs and legislations such as the EU Water Framework directive, the UN Millennium Development Goals and the Kyoto Protocol (WETwin 2008).

18.1.2 Justification of the Research

In Africa and Latin America wetlands have been threatened because of unsustainable management (WETwin 2008). Among many causes for the loss of wetlands, Wang (2008) presents the following list: lack of public awareness, insufficient funding, imperfect legal system, insufficient wetland research, lack of coordination among agencies and unclear responsibilities and undeveloped technologies.

Wetlands, because of their value and ecological richness are the focus of many dilemmas as they, on the one hand need to be conserved and at the same time fulfill users' requirements. Hence, management can become very complex, and many decisions about tradeoffs have to be taken. Consequently, the need to comprehend the factors that influence decision making and the role of national and international guidelines in this process is particularly appealing to better understand the management system.

According to the Ramsar Convention, putting in place a sustainable wetland management system is an essential tool to achieve the wise use of wetlands. Years ago it was believed that wetland management could be resolved with extensive and accurate information, and nowadays this has been determined with the availability of technology; but still we are loosing wetlands. Recently, the concept of participation has been adopted as an option, but the practice reveals that this is not applied everywhere. This research focuses on the decision-making process in order to better understand the main cause of wetland deterioration in developing countries, including Ecuador.

18.2 Methodology

The outline of RBM developed by Mostert et al. (1999) was used as the basis for the conceptual framework for the assessment of wetland management. Some changes and adaptations from Mostert's outline to this framework were made, mostly

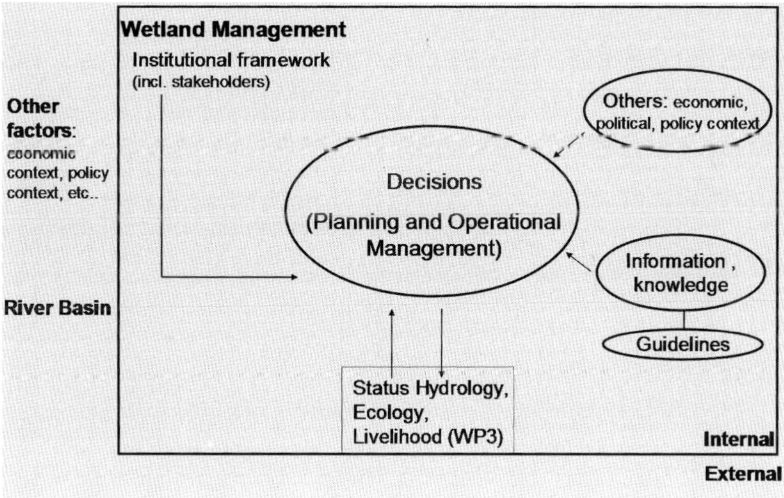

Fig. 18.1 Conceptual framework

because other factors were missing, such as the influence from economy, political context and policy.

In this conceptual framework there is the outside influence of the River Basin, due to the importance of the relationship between water and land, pointing out the geographical situation (downstream – upstream); Mostert et al. (1999). Additionally, the importance of decision making has a central role, considering a central process within a complex context such as wetland management (Fig. 18.1).

Using the Conceptual Framework, for practical purposes an Analytical Framework was developed. This last one consists of seven parts; the ones that related to the conceptual framework are shown in Table 18.1. This analytical framework was implemented in Ecuador for the case of Abras de Mantequilla wetland.

18.2.1 Case Study: Abras de Mantequilla Wetland: Los Ríos Province/Ecuador

In this case the objective was to analyze how things were done in wetland management and more specific, how decision were made, being of primary importance to define some relationships. Abras de Mantequilla wetland has an extension of about 65,000 ha and is located in Guayas River Basin, which is very important for Ecuador, due to their agricultural activities (Fig. 18.2).

The following list is the criteria by which this particular wetland was chosen. The reason for the selection related to the opportunity and accessibility of information, and relevance of the site at local, national and international level (Table 18.2).

Table 18.1 Linkages between the conceptual and analytical framework

Conceptual framework	Analytical framework
Influence of the river basin, "other factors"	1. Description river basin
General context of the wetland, "others"	2. Description wetland
Wetland management (Institutional framework)	3. Management structure of the wetland
Information – knowledge – guidelines	4. Wetland guidelines and indicators
Decisions in the planning process	5. Wetland planning in practice
Decisions in the operational management process	6. Wetland operational management in practice
Relation river basin – wetland	7. Adaptive capacity

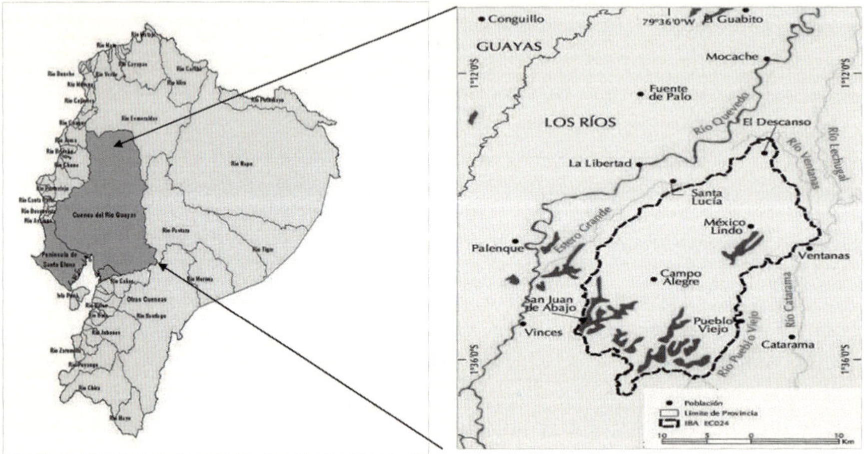

Fig. 18.2 *Left*: location of the wetland River Basin in Ecuador. *Right*: location of the wetland in the Guayas River Basin

18.3 Results

In this section the results of the fieldwork in Abras de Mantequilla are presented. In the following Table 18.3, there is a list of the factors that have been identified by the interviewees, and the importance they give to each of them. This will help with the comprehension of how decisions are made inside the wetland and in relation to it.

1. **Political Factor**: Ecuador's population is very engaged in politics, which is the reason why people mention it as a separate factor. Moreover, in reality the other factors are very much interconnected with politics.
   The economical, social and environmental factors happen in a political scenario; in conclusion, first the site has to wait to be part of the national political agenda. Therefore, it will gain importance if it becomes a political resource for acquiring votes, money and power. Secondly, after becoming an asset, the site can start the application of Ramsar and national requirements to achieve a wise use of the wetland. The Ministry of Environment, RBO (River Basin Organization) and Commonwealth, because of their status as public entities, have a political duty; therefore, politics is the first factor that is taken into account when making a decision.

Table 18.2 Criteria for case study selection

	Abras de Mantequilla		
	Criteria		Reasons
O/A	Available information	√	Internet sites were the first source of information and a very accurate one. In the country printed documents were available
RQ	Conserve natural conditions	√	This is a 1–4 and 6–8 Ramsar site, meaning that it is representative because of its habitat and large area, it has not being altered by human activity maintaining the biodiversity of the habitat
RQ	Population's dependency on the wetland	√	The main activities of the people around the wetland depend on it: agriculture and fisheries
RQ	Wetland management in place	X	This is not in place, but the ministry of environment is very involved in the area and there is a pilot plan recently developed, showing a form of organization
RQ	Part of an organized river basin	√	The wetland is part of the biggest and most important River Basin in the country: the Guayas River Basin, and it has a RBO (River Basin Organization) in place
RQ	Ramsar site	√	This wetland is consider of international importance since 2000
O/A	Contacts	√	The counterpart of the WETwin project in Ecuador had already the knowledge of the key contacts in the area
O/A	Communication facilities	√	Spanish is the language in the country and internet facilities are very widespread in the country, even in rural areas

RQ Research questions, *O/A* Opportunity/Accessibility

Table 18.3 Factor and actors in the decision making process

	Actors				
Factors	Ministry	River basin organization	Commonwealth	Locals	Inhabitants
Politic	1	1	1	2	3
Economic	4	2	2	1	1
Environmental	2	3	4	3	4
Social	3	4	3	2	2

The order of Importance from: 1 the most important and 4 the less important factor

2. **Economic Factor**: This case study was conducted in a developing country; the State cannot cover all the expenses of the public institutions from the central budget. Therefore, the RBO and the Commonwealth have to find other ways to finance their works, which is why this factor is in second place of importance to influence their decision-making process.

   For the locals (people living in the region, not specifically in the wetland) the economical issue is of primary importance. Therefore, any decision they made is related to the perceived benefits for the improvement of their lives: development, more work, and income.

   Finally, inhabitants of the wetland based their decisions on whether their basic needs such as food and shelter will be covered; therefore, they will lean towards the option that will assure their economic situation.

3. **Social Factor**: This community has proved to be very active and willing to fight against anything that can jeopardize their wellbeing. This has been shown through the protests against informal garbage dumping in the wetland, where an informal social movement was organized and supported by local and national authorities. The response from society to specific actions has a predominant value in the process of decision making. Local institutions, as the Commonwealth, are aware of the power social movements can have; consequently they consider this factor as more important compared to other national institutions.
4. **Environmental Factor**: One might assume that the environment would be the first factor considered in the decision-making process due to the Ramsar status. To the contrary, it is the last factor to be taken into account, due to other needs that have to be addressed first by stakeholders and decision makers.
   Moreover, problems concerning the environment are not obvious to the naked eye, which is the reason why outsiders (decision makers) overlooked this factor. It seems that water is sufficient and clean, animals are present and people look healthy around it.

18.3.1 The Role of Ramsar

Abras de Mantequilla has been a Ramsar site for almost 10 years now. In 2008 the authorities, noticing the lack of applicability of this Convention, made an evaluation. The results showed that Ramsar has been used mostly to fill the legal emptiness that presented the environmental law on wetlands, and their handbooks utilized for the development of plans.

In the case of Abras de Mantequilla, there is a Pilot Plan, which used the handbook about management planning, specifically for the action plan that was proposed. The central office of the Ministry of the Environment is aware of the poor diffusion of Ramsar guidelines, but also blames the people for their lack of interest in research.

Furthermore, the statements the Convention presents in their guidelines center the discussion on the environment. For this reason the country has noticed the need to interact with other institutions to define the applicability of Ramsar guidelines, in order to have a wider point of view (social and economical primarily).

Finally, in Ecuador all Ramsar guidelines are perceived as the most trustworthy documents that exist. The local environmental authorities do not see the need to invest time in developing more guidelines; to the contrary, they see the need for finding ways to apply them to local realities.

18.4 Conclusions

Wetlands can be the solution to different problems, and most of the time decision makers overlooked the complexity of its management. Through this case study the different factors that push forward the decision-making process have been presented.

Furthermore, the importance these actors give to each factor is what shaped the future of the ecosystem and its functions.

In conclusion, the interests of the most influential institution or actors are the ones that will be taken into account in the decision-making process. Those considered the strongest can be decision makers or stakeholders, depending on how much noisc and pressure they have over the population and resources.

Consequently, it becomes a game of powers: the actor which has the potential to cause the biggest losses due to their actions, or has the power to produce considerable pressures (in an economical, political, social or environmental way) is the winner; and their priorities (their factors) will lead the process for making a decision.

References

Mostert E et al (1999) River basin management. In: Proceedings the international workshop, The Hague, pp 27–29

Wang Y et al (2008) Wise use of wetlands: current state of protection and utilization of Chinese wetlands and recommendations for improvement. Environ Manage 41:793–808

WETwin (2008) Enhancing the role of wetlands in IWRM for twinned river basins in EU, Africa and South America in support of EU Water Initiatives

Chapter 19
3D Detrital Aquifer Modelling for Water Resources Management of the Regional Park of the Lower Courses of Manzanares and Jarama Rivers (Madrid, Spain)

F. Carreño-Conde, S. García-Martínez, J. Lillo-Ramos, and R. Fernández-Martínez

Abstract Water resources management of natural sites requires a powerful tool to analyze the process and changes that are occurring in the environment. This paper describes a 3D geo-model design of the Jarama River Detrital Aquifer located in Madrid (Spain). The hydrogeological unit is included in the "Parque Regional de los Cursos Bajos de los Ríos Manzanares y Jarama" (Regional Park of the Lower Courses of Manzanares and Jarama Rivers). All data used in this study were integrated in a geographic database (GDB): geological and hydrogeological information, geological map (1:25,000), 11 cross-sections, isopieces maps and a Digital Elevation Model (DEM). The constructed 3D model of the Jarama Aquifer shows the geometric features and the spatial distribution and variation of geologic units, which allow the assessment of volumes of each unit, the depth and thickness variations of the main aquifer, and the spatial and temporal variations of water tables.

Keywords Water resources management • Geo-model • Geographic database • Aquifer modelling • Madrid • Spain

F. Carreño-Conde (✉) • J. Lillo-Ramos
Dpto. de Biology and Geology, University of Rey Juan Carlos, C/Tulipán s/n, 28933 Madrid, Spain

Madrid Institute of Advanced Studies in Water Technologies IMDEA Water, C/Punto Net 4, 28805 Alcalá de Henares, Madrid, Spain
e-mail: francisco.carreno@urjc.es

S. García-Martínez • R. Fernández-Martínez
Dpto. de Biology and Geology, University of Rey Juan Carlos, C/Tulipán s/n, 28933 Madrid, Spain

H. Farfán González et al. (eds.), *Management of Water Resources in Protected Areas*, Environmental Earth Sciences, DOI 10.1007/978-3-642-16330-2_19,

19.1 Introduction

Three-dimensional (3D) reconstruction techniques for the study and analysis of complex geological bodies are mainly used in hydrocarbon exploration and production (Perrin et al. 2005), analysis of structural geology (Martelet et al. 2004), reconstruction of geological surfaces (Fernández et al. 2004), mining exploration (Le Carlier et al. 2009; Feltrin et al. 2009), environmental risk (Wycisk et al. 2009), or hydrogeological modelling for groundwater resource management (Bonomi 2009; Gallerani and De Donatis 2009). That is because with a 3D reconstruction a consistent model can be obtained that describes a detailed succession of different layers with their geometrical relationships and spatial distribution, allowing the calculation of the volume of each layer. Also, the 3D true representation allows a best understanding of the complex subsurface settings of different units (Nury et al. 2009).

Groundwater resource evaluation and management require the adequate spatial representation of geological and hydrogeological information including that regarding rock layers and water table depth, in order to obtain isodepths and isopiezometric maps, vertical cross-sections, and geometric solutions of the piezometric surface of the aquifers, closed volumes and data estimation of groundwater reserves in the aquifer of interest.

The aim of this work is to show how cartographic, topographic, geological and hydrogeological data can be integrated to generate a 3D geo-model of the Jarama River Detrital Aquifer, giving as a result a powerful basis for the analysis of changes that occur in the territory that must be taken into account in making a decision regarding the environment management. This aquifer is included in the "Parque Regional en torno a los ejes de los cursos bajos de los ríos Manzanares y Jarama" (*Regional Park of the Lower Courses of Manzanares and Jarama Rivers*). This is located in Madrid (Spain) and coincide with a site of community importance (SCI) code SCI-ES-3110006 "Vegas, cuestas y páramos del Sureste" (Fig. 19.1).

There is a clear relationship between the natural resources, ecosystems, activity in the territory (agriculture or mining) and the groundwater flow systems. The natural equilibrium established can seriously change if there is an increase in the intensive use of elements.

19.2 Data and Methodology

All data used in this study are stored in a geographic database (GDB) edited in ArcGIS 9.2. for the management of geological and hydrogeological information. A 3D Core Builder (3DCB) extension for ArcView 3.x. (O'Neall 1999) with 3D Analyst was used for making 3D lithology cylinders (boreholes) and 3D Analyst extension for ArcGIS 9.2 allowed true surface interpolations. ArcScene application integrated witch ArcGIS was selected for 3D visualization because it allows the view of GIS data in three dimensions.

The main data source for this work is the "Estudio para la ordenación de la actividad extractiva en el tramo bajo del río Jarama" (*Study for managing the extraction activity*

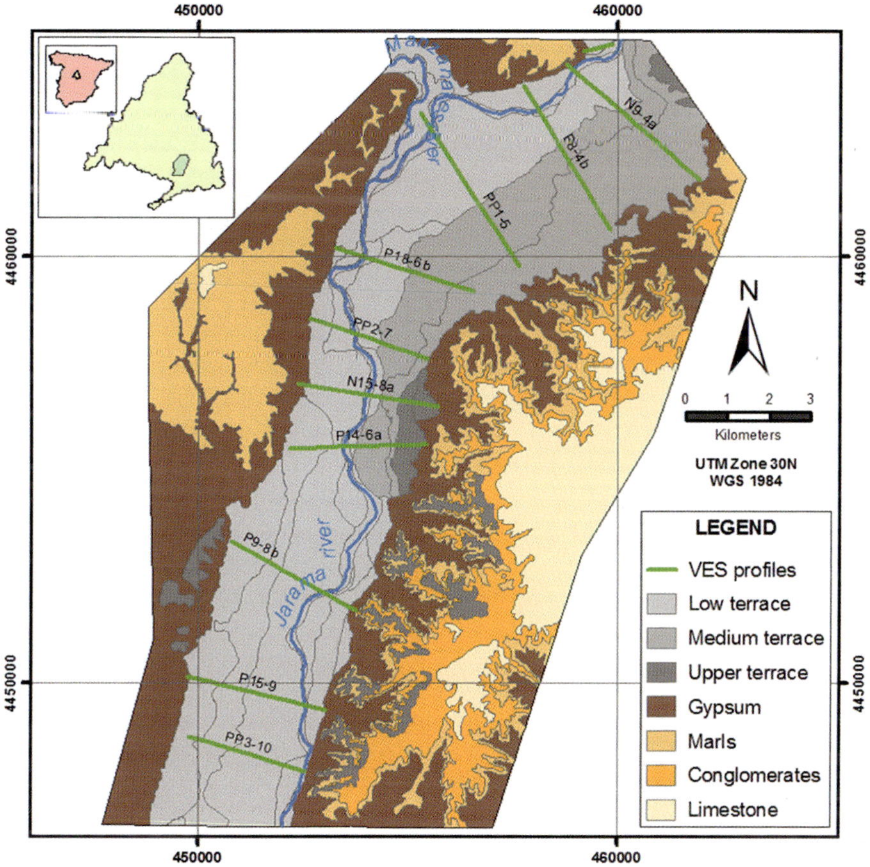

Fig. 19.1 Location map of the study area with geologic units considered for the 3D model. Location of the VES (vertical electrical soundings) profiles is also shown

in the lower Jarama river) (Bardají et al. 1990) that provide a geological map (1:25,000), 11 cross-sections obtained from vertical electrical soundings (VES) interpretation and isopiezometric maps of the Jarama River Detrital Aquifer. Additionally, a Digital Elevation Model (DEM) was obtained for surface topography of 25 m resolution and four boreholes from Tajo Hydrographical Confederation database.

The procedure for producing a 3D model can be summarized as follows. Firstly, the collected information is prepared and imported according to a geodatabase structure in the UTM reference system. The previously georeferenced geological map is digitalized on screen. Afterwards, the geological boundaries are converted to polygon geometries with attributes that represent the spatial distribution of lithological-hydrogeological units. They consist of sediments (limestone, conglomerates, marls and gypsum) of Tertiary age and Quaternary sediments including those materials eroded, transported and deposited by the Jarama and Manzanares rivers. These sediments are conforming the upper, medium and lower river terraces, and the alluvial filling of the valley bottom. The 11 cross-sections distributed along the study area were converted in virtual 3D boreholes that contain the sediments vertical distributed each 100 m.

The next step was the construction of the considered lithologic surfaces. Using a new point layer with 79,295 elements arranged in 50 m intervals the study area is regularly discretized to obtain virtual 3D lithologic boreholes. Each borehole contains upper and lower topographic heights of geologic units. The built 3D model is based on a constructive method so that each model surface is processed independently, assuming that all layers are characterized by a horizontal disposition and constant thickness. The upper layer height is the same as topographic height of MDE and the lower layer height is derived from minimum topographic height of the layer boundary. In the points that do not match with emerging surface, the upper layer height is the same that lower height of overlaying layer. The gypsum unit is considered as the impermeable layer of Jarama Detrital Aquifer, so the design of its subsurface is crucial in the model. However, the gypsum unit displays an irregular pattern and its surface can be obtained from the geologic map. Alternatively, their surface was interpolated from those depths extracted of the 11 VES cross-sections.

The isopiezometric curves were converted to a point layer in order to build the piezometric surfaces of the unconfined Jarama Aquifer using inverse distance weighted (IDW) interpolation.

Afterwards, the geologic surfaces must be interlocked to validate the results. A careful checking of geological settings, virtual 3D boreholes and surface topography was performed to detect inconsistencies in the conceptual geological model and the data sources. Thus, the revised 3D model contains an individual representation of each formation, and a complete 3D model of the Jarama Aquifer is generated when they are overlaid.

19.3 Results and Discussion

A 3D geological model was obtained using a variety of geological, hydrogeological and topographic data. The 3D model of the Jarama Aquifer shows the spatial distribution and geometric features and variability of geologic units. This design allows the displaying of all units as a detailed succession of layers or alternatively, the model allows working with each formation independently.

Four layers constitute the obtained 3D model of the Jarama Aquifer, defined by lithological and hydrogeological criteria with an environmental significance: layer 1 is composed of gypsum of Middle Miocene age, occurring along both margins of Jarama river; layer 2 consists of marls, conglomerates and limestone deposits of Middle to Upper Miocene age occurring in the western section of the Jarama valley; layer 3 is composed of gravels and alluvial deposits of Quaternary age; and layer 4 is made up of limestone, marls, conglomerates and terrace deposits (Fig. 19.2).

This study is focused on the layer 3 and layer 1. On the one hand, layer 1 consists of gypsum and is considered the basement of the Jarama aquifer system, defining a very low permeable formation and a natural barrier that confines and prevents a

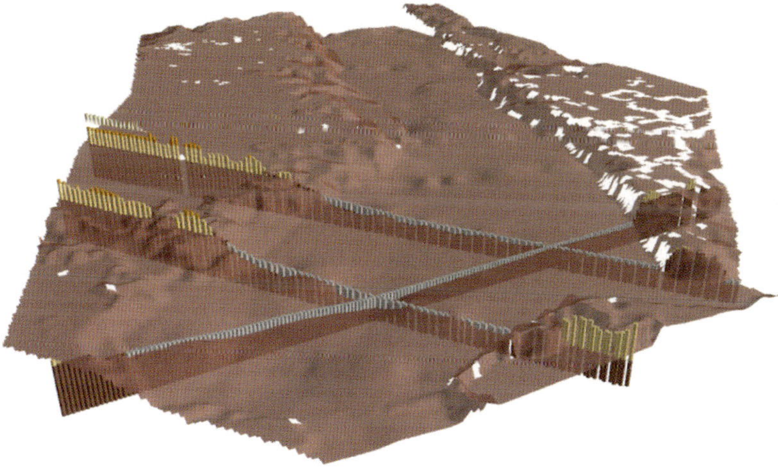

Fig. 19.2 Representation of the real 3D model of the area studied with boreholes cross-sections and a layer contact surface. Each considered layer is differentiated by a colour: Layer 1 *brown*; Layer 2 *yellow*; Layer 3 *orange*; and Layer 4 *gray*

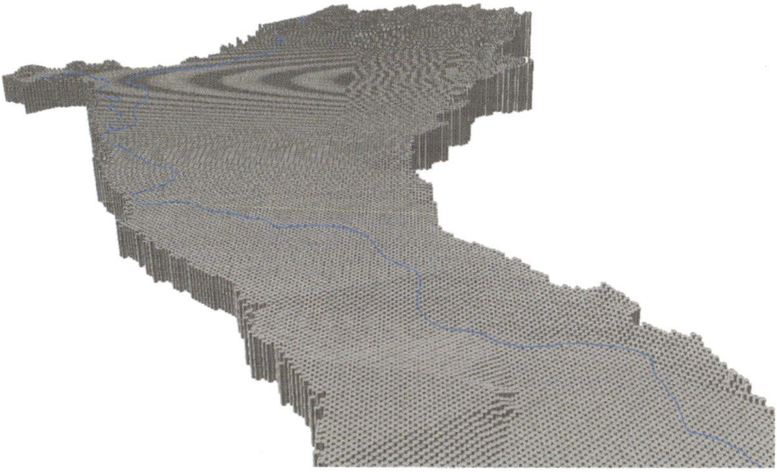

Fig. 19.3 Visualization of the 3D main Jarama aquifer system model (layer 3 gravels) with Manzanares and Jarama rivers

possible transport or dispersion of a potential contaminant plume. The defined upper gypsum surface is similar to that outlined in the VES cross-sections and its lateral continuity is coherent with surface geological map. Layer 3, composed of gravels, is considered the main aquifer; therefore, it is very important to know their geometry and spatial boundaries for groundwater management (Fig. 19.3).

Moreover, if this unit is affected by extractive industries (gravels) the management should be sustainable by law.

The results are correct and coherent with the general geological knowledge but the local spatial variations of the Tertiary and Quaternary sediments or the human activities could not be considered in the model, to obtain a more accurate environmental setting. A "true" model depends on density and availability of information according to scale (Wycisk et al. 2009). As more data are involved, models are more detailed.

In this model, the unconfined Jarama Aquifer volume is estimated as 1.04×10^{10} m^3. The created isopachs map shows a thickness variation of the main aquifer from 1 to 56 m, with the maximum depths located to the northeastern and southeastern parts of the study area.

Analysis of the isopiezometric models obtained from 1973 to 1989 shows a light increase of the water table depth (mean of 0.39 m), being the estimated average volume is 3.7×10^9 m^3. The maximum increases are located along the aquifer and the main decreases at the northwestern and the southern parts of the boundary.

19.4 Conclusions

A 3D model of Jarama Aquifer built using GIS techniques based on geological and hydrogeological data can help to better understand the complex subsurface setting because it enhances the visual interpretation of the sedimentary layers that form the aquifer.

The model is based on a constructive method that uses a variety of geological, hydrogeological and topographic data from geologic maps, boreholes, cross-sections, DTM and outcrops for which integration allows the assessment of the geometry and spatial variability and the volume of the sedimentary layers and water storages.

The piezometric surface is generally near the surface zones and in most of the cases a hydraulic connection exists among the aquifer, the river and the wetlands. The intersection between the main aquifer isopach and isopiezometric models can help to establish the potential risks derived from an intense groundwater exploitation rate, extractive mining activity or a groundwater contamination and their effects, which is paramount for the management and conservation of the regional natural park. In addition, the model can also be used to analyse the different scenarios for natural recharge, and to use planning of water and environmental resources.

The Jarama Aquifer 3D model represents a key tool to estimate the natural resources, environmental risk and potential negative impact derived for anthropogenic activities. All of it is required for a correct management of the regional park, allowing the application of regulations to mitigate the negative effects of human activities.

Acknowledgments The authors wish to thank María Bascones and Luis Toribio from the Directorate General for the Environment, Department of Water Quality of the Community of Madrid for his support and the information. They also wish to thank Mercedes Echegaray (Hydrographic Confederation of Tagus River) for his comments and suggestions.

References

Bardají I, Cabra P, Calvo JP et al (1990) Estudio para la ordenación de la actividad extractiva en el tramo bajo del río Jarama. Servicio de Estudios y Planificación. Agencia de Medio Ambiente. Comunidad de Madrid

Bonomi T (2009) Database development and 3D modelling of textural variations in heterogeneous, unconsolidated aquifer media: Application to the Milan plain. Comput Geosci 35:134–145

Feltrin L, McLellan JG, Oliver NHS (2009) Modelling the giant, Zn–Pb–Ag century deposit, Queensland, Australia. Comput Geosci 35:108–133

Fernández O, Muñoz JA, Arbués P et al (2004) Three-dimensional reconstruction of geological surfaces: an example of growth strata and turbidite systems from the Ainsa basin (Pyrenees, Spain). Am Association Petroleum Geol Bull 88(8):1049–1068

Gallerini G, De Donatis M (2009) 3D modeling using geognostic data: the case of the low valley of Foglia river (Italy). Comput Geosci 35:146–164

Le Carlier de Veslud Ch, Cuney M, Lorilleux G et al (2009) 3D modeling of uranium-bearing solution-collapse breccias in Proterozoic sandstones (Athabasca Basin, Canada) Metallogenic interpretations. Comput Geosci 35:92–107

Martelet G, Calcagno P, Gumiaux C et al (2004) Integrated 3D geophysical and geological modelling of the Hercynian Suture Zone in the Champtoceaux area (south Brittany, France). Tectonophysics 352:117–128

Nury S, Zhu X, Cartwright I, Ailleres L (2009) Aquifer visualization for sustainable water management. Manag Environ Qual Int J 21(2):253–274

O'Neall M (1999) User manual 3D core builder. Version 3a. Indian Geological Survey

Perrin M, Zhu Z, Rainaud JF, Schneider S (2005) Knowledge-driven applications for geological modelling. J Petroleum Sci Eng 47:89–104

Wycisk P, Hubert T, Gossel W, Neumann C (2009) High-resolution 3D spatial modelling of complex geological structures for an environmental risk assessment of abundant mining and industrial megasites. Comput Geosci 35:165–182

Part V
Research and Monitoring of Water Resources in Protected Areas

Chapter 20
Preliminary Characterization of the Hydrodynamic of Karstic Aquifer in the "Promontorio de Cabrera", Province of Maria Trinidad Sánchez, Dominican Republic

E. Rocamora-Álvarez

Abstract The "Promontorio de Cabrera" is represented by a mountainous massif of 250 km^2, with elevations of more than 400 m of height, distributed to 6 km from the coastline, and conformed by karstic limestone rocks. There is an aquifer with conical structure and radial flow that responds to the morphological regional patterns, and where the groundwater has depths of 200 m in the central sector of the massif and of 80 m in the borders. The aquifer structure has a better component developed in NW, N and NE, where the groundwater appears in karstic springs, which have altitude of 120 and 180 m and flow to the sea. Other springs in the NNE sector of the massif are distributed to heights of 280 m. In the S and SE of the massif, the karstic springs are distributed to heights of 120 and 150 m, and the flow is to the local river and wetlands. Other two springs, in the level of 260 m, suggest its hydrogeology relationship with the morphological structure. A shallow aquifer associated with the system of karstic summits has been characterized, where the waters appear in karstic springs and flow to basins and infiltrate for karstics drains.

Keywords Karst aquifer • Karst springs • Hydrogeology • Morphology • Dominican Republic

20.1 Introduction

The territory described as the study zone is located in the province of Maria Trinidad Sanchez, in the north coast of Dominican Republic, embracing an area of 25 km^2 (Fig. 20.1). This territory is represented by a mountainous massif, where

E. Rocamora-Álvarez (✉)
Geophysics and Astronomy Institute, Calle 212 No. 2906 e/29 y 31, La Coronela, La Lisa, CP 11 600 Ciudad de La Habana, Cuba
e-mail: rocamora@iga.cu

H. Farfán González et al. (eds.), *Management of Water Resources in Protected Areas*, Environmental Earth Sciences, DOI 10.1007/978-3-642-16330-2_20,

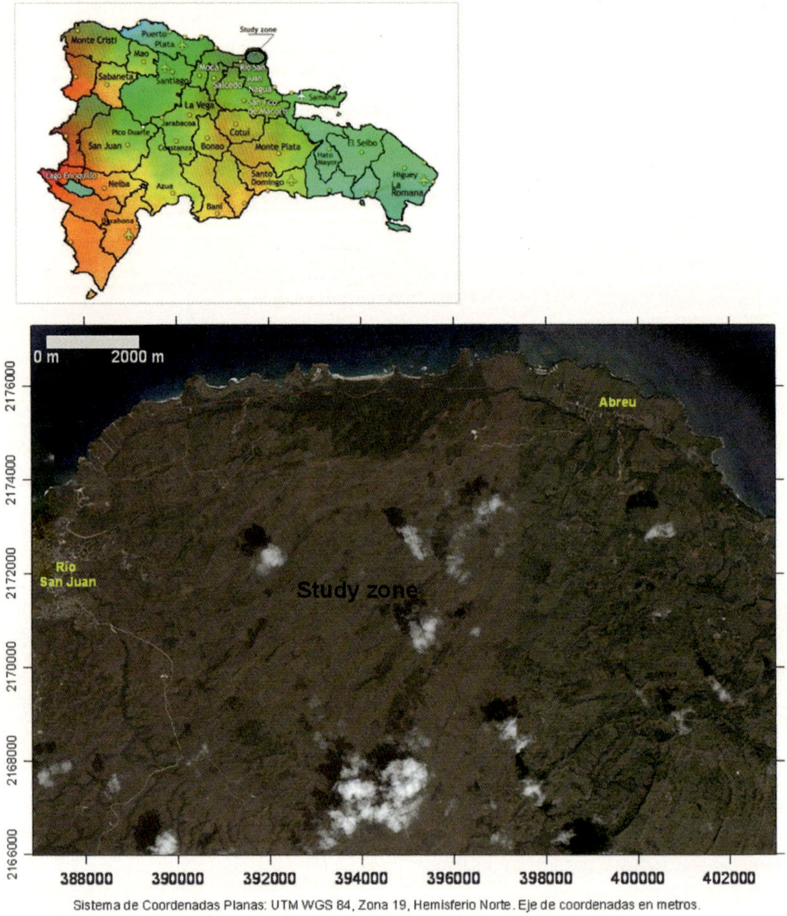

Fig. 20.1 Study zone location

to less than 10 km from the coast there are elevations of 400 m, and the karst in the carbonated rocks established the presence of the groundwater.

The basic characterization of the hydrogeological and hydrological aspects was based on the evaluation of the local morphology, of the rocks distribution, of the karst development, of the physical and chemical parameters of the groundwaters that appear in karstic spring and of the ephemeral hydrological system.

20.2 Geologic and Geomorphologic Characteristics

The stratigraphic sequence is evident in the prevalence of the carbonated complexes that distributed over rock layers of the regional metamorphism (Direccion Geneal de Minería 1999, 2000).

From a local point of view, this stratigraphy is simplified to:

- *Limestone, karstics and fractured, of white color*. These calcareous rocks are of the Miocene-Pliocene period. They are hard rocks, where the karst has been developed following the patterns of the primary cracking of the massif
- *Loams to calcareous marl, white to cream, hard and dense*. These rocks lie in layers inserted in the limestone, but with a limited development inside the study area.

The evaluation of the natural density of the rocks approached the interpretation of the rocks distributions, in the intercalations of loams and limestone, distributed with general strata azimuth of SW-NE.

The regional geomorphology becomes interesting when considering that the whole territory is framed inside the limits of "Promontorio de Cabrera", like a geomorphologic unit and surrounded elements of coastal and fluvial plains.

In the study zone are observed some levels of morphostructural surfaces. One of these surfaces is characterized as I that it is of more altitude in the region, with elevations above 280–300 m and it occupies 13 % of the total area of the massif. Its land surface is very irregular, with summits with patterns of alignment of 30–40°.

Surface II is distributed between the elevations of 100 and 260 m. This is the surface of major extension in the territory (112 km^2). It is very irregular, where there are the summits arrays aligned with direction 30–45° and deep valleys of vertical hillsides with its bottoms filler for sediments and with active drains. Some of these valleys represent closed basins of small dimensions.

Associated with its surface there are several springs of groundwater that have a similar pattern of direction, of distribution and elevations. These springs develop into the fluvial stream of the territory, with permanent and seasonal flow.

Surface III is distributed from 0 to 30 m, around the massif and it is the "last" level of emerged terrace. It represents the coast border with flat relief and covered with marine and alluvial sediments.

Another factor refers to the interpretation of the morphostructures direction in the study zone. Of this interpretation three families of morphostructures have established: the family A with directions 30–60° that it represents the direction of distribution of summits, valleys and structural steps.

Family B contains the structures with directions 130–160°, with a considerable continuity. These alignments represent borders of blocks and relief structures, without contributing to the superficial and underground hydrodynamics.

Family C has a very small representativeness in the massif, with directions 90–100°, that it represents morphostructures of reduced continuity and null aqueousness.

20.3 Aspects of Hydrology and Hydrogeology

Due to the karst development on the study zone, the superficial streams do not flow. This condition establishes that the superficial hydrology is manifested in three different ways:

- *Elevated sector of the mountainous massif*. Valley that functions as small closed basins, where the pluvial flow to the bottom; and they infiltrate for karstic drains.

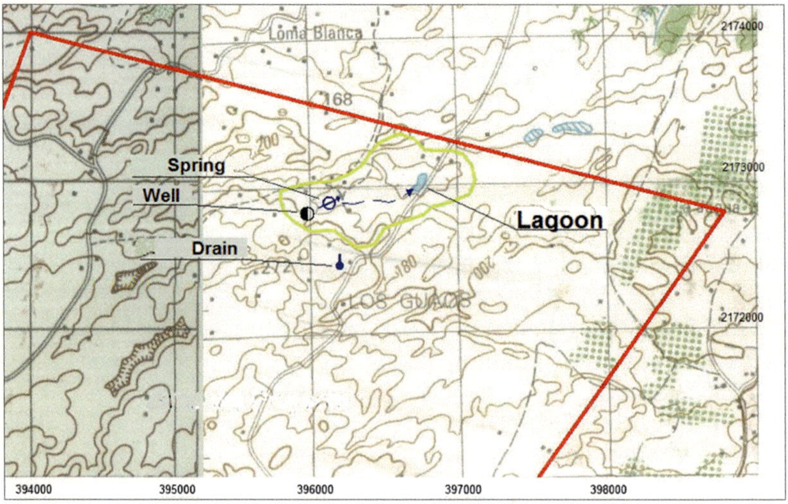

Fig. 20.2 Elements that characterized the study basin

This dynamic has an eventual character, although they are observed in some seasonal streams.

- *Plain of border on the North and East sectors of the massif.* There are permanent fluvial streams, with continuous riverbeds. The distribution of these streams and their predominant directions follow the patterns of directions of the family A morphologies.
- *Hydrological basin of the "Caño Claro" stream.* This basin is distributed in the South flank of the mountainous massif and the fluvial streams that flow to the basin appear in the border of massif. These streams are fed by the pluvial from the high sector of massif and from aquifer.

The hydrology of the central sector of the massif is represented by a system of closed basins in a karstic environment of very irregular surface that constitutes the area of infiltration of the pluvial waters. This zone is defined as the area of feeding of the aquifer.

During the times of strong rains some water streams appear to flow from de massif to the plains, but the infiltration process is predominant. These groundwaters represent the karstic aquifer, with radial flow whose piezometric level is more than 80 m of depth. Due to the massif morphology these groundwaters appear at elevations between 120 and 170 m, as these are the level of regional hydrology. From these morphology levels and slope below, these streams have the permanent or seasonal hydrodynamic to discharge to the sea.

However, this conceptual hydrogeologic model is represented by another dynamic of the groundwaters that flow near the massif surface, and its movement is through the karstic hollows (León and Rocamora 1996; Molerio et al. 1998).

In the map of Fig. 20.2 is represented this hydrodynamic in one of the closed basins, with the preferential direction characterized for these relief forms (60° – family A).

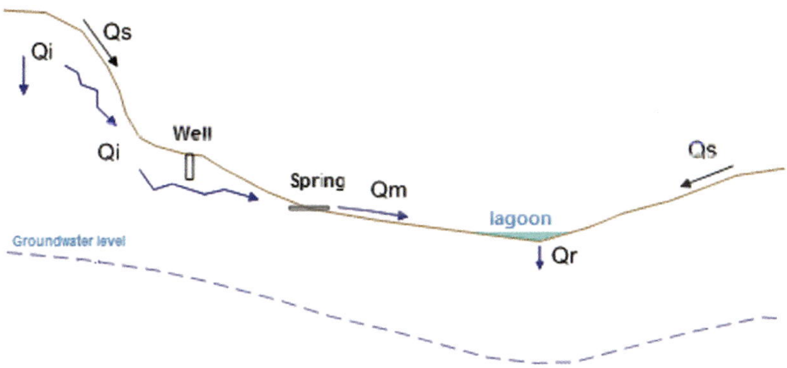

Fig. 20.3 Hydrodynamic profile in the study basin for the dry period

The well has a vertical orientation of square section and its depth is about 3.00–3.50 m. It was dry during the field campaign and elevation of its opening is 193 m. On the slope is the water spring and it has 190 m of elevations and there is a lagoon in the bottom of the basin where are accumulated the surface waters (elevations is 177 m). The drain corresponds with the near basin and its altitude is 224 m.

The hydrodynamics in the basin behave in two ways, in dependence of the meteorological conditions of wet period or dry period:

The dry period (Fig 20.3). This situation is considered as the least favorable for the local hydrology, without it means the conditions of the period of drought.

The scarce limited rains produce the reductions of pluvial volume (Qs) and so the volumes that infiltrate to the massif (Qi). This situation influences directly in the massif aqueousness that is reflected in a decreasing of the flows to the deep aquifer (Qas) and also on the shallow waters that flow in the unsaturated area of the massif. These conditions established the water flow in the spring Qm, with a tendency to reduce if the conditions of insufficient rain persist.

The conditions of accumulated waters are keeping in the lagoon, although there exists an infiltration flow Qr.

The wet period (Fig. 20.4). This situation is favorable for the local hydrology, although some conditions of floods are observed.

The increase of superficial water flow (Qs) and its infiltration (Qi) influence directly the aquifer recharge, that which increases the intrinsic aqueousness of the karstic massif and the water flow of the spring (Qm).

These conditions produce the increase in volumes of water that are accumulated in the lagoon and also the increment of the infiltration flow Qr.

Anyone of these situations represent the hydrodynamics of the superficial and groundwaters in the basins of the mountainous massif and toward the valleys.

Considering these approaches, a general analysis of the water quality was made, taking as a reference the physical-chemical parameters, measured in situ (Table 20.1).

In general, the waters were sampled in the karstic springs and in the wells have a very short time of residence in the massif, demonstrated by its low values of SPC

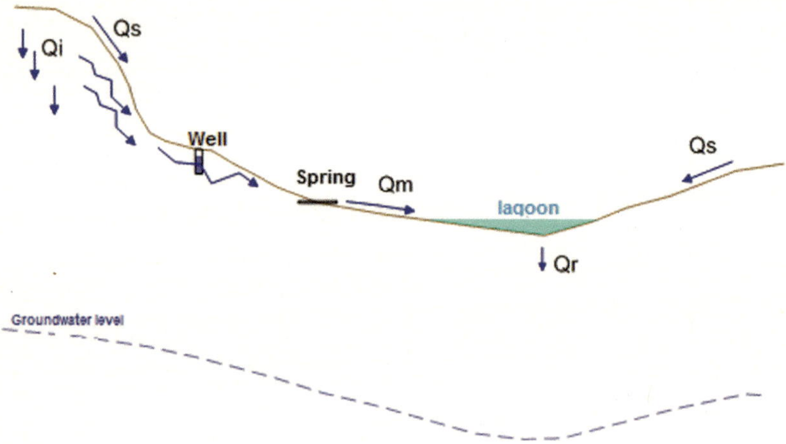

Fig. 20.4 Hydrodynamic profile in the study basin for the wet period

Table 20.1 The physical-chemical parameters of the superficial and groundwater

Sampling points	pH	t °C	SPC (μS/cm)	TDS (mg/l)	O_2 (mg/l)
M1	6.80	26.0	540	262	7.33 (98.1 %)
M2	7.09	26.0	346	167	5.03 (87.2 %)
M3	6.54	27.3	642	311	6.23 (94.1 %)
M4	7.12	27.5	441	214	4.23 (83.2 %)
M5	6.77	27.6	467	226	4.66 (85.2 %)
M6	7.02	27.8	398	193	5.20 (89.4 %)
Well 1	7.02	22.0	678	329	7.33 (98.1 %)
Well 2	7.09	22.6	632	306	7.03 (97.2 %)
Well 3	7.04	23.1	589	286	4.23 (84.7 %)

t temperature (°C), *SPC* conductivity (μS/cm), *TDS* total dissolved solids (mg/l), O_2 dissolved oxygen (mg/l) (% of saturation)

and TDS. Also, this characteristic demonstrates the rapid infiltration and migrations of water in the karstic massif.

Considering these parameters, the insignificant difference of these properties for the waters of the deep aquifer and the shallow can be observed.

Acknowledgments Thanks to the executives of the Enterprises EMPACA REDES of Dominican Republic for their support in the field investigations

References

Dirección General de Minería (1999) Mapa Geológico de la República Dominicana a escala 1: 250,000

Dirección General de Minería (2000) Proyecto Programa de Desarrollo Geológico Minero (SYSMIN)

León A, Rocamora E (1996) Evaluación del riesgo de contaminación del acuífero por emisiones de gas licuado de petróleo en la refinería "Ñico López", Ciudad de la Habana. Technician report of the Center of Hydrology and Waters Quality, INRH, Cuba

Molerio LF, Guerra MG, Torres JC, Rocamora E (1998) Vulnerabilidad Intrínseca y Riesgos Ecohidrológicos del Olcoducto Varadero-Matanzas. Technician report of the Group of Terrestrial Waters, IGA-CITMA, Cuba

Chapter 21
Underground Water of Deep Circulation in the National Park Guanahacabibes, Pinar del Rio Province, Cuba: Another Alternative with Water Supply Aims

R. Peláez García and N.A. González Cabrera

Abstract The National Park Guanahacabibes, located in the westernmost part of Pinar del Rio province, occupies an area of 1,200 km^2. It was recognized in 1987 in an official way and internationally as a town of biospheric transcendency. From the geologic point of view it is composed by Neogene – Quaternary rocks over other older Jurassic rocks with a great development of the karst of the plain that limits the existence of superficial fluvial flows. In those carbonated rocks, there is an aquifer open to the sea and strongly intrusive by marine waters, and the only possibility of finding sweet waters in the small lenses that float on the salted waters. As an alternative related to the different proposals carried out by the authors in different moments with the objective of solving the water supply to the peninsula, in this work there is an approach to the possibility of finding sweet waters of deep circulation in older rocks (Jurassic), outlining a Conceptual Model Geologic – Hydrogeologic Regional, using current interpretations.

Keywords Water supply • Karst • Aquifer • National Park Guanahacabibes • Cuba

---

R. Peláez García (✉)
Geominera Enterprise Pinar del Río, MINBAS, Pinar del Río, Cuba

Department of Geology, University Hermanos Saiz, Pinar del Rio, Cuba
e-mail: bpinar@pri.minbas.cu

N.A. González Cabrera
Department of Geology, University Hermanos Saiz, Pinar del Rio, Cuba

Research Center and Environmental Surveys ECOVIDA-CITMA, Pinar del Rio, Cuba
e-mail: nelson@ecovida.vega.inf.cu

H. Farfán González et al. (eds.), *Management of Water Resources in Protected Areas*, Environmental Earth Sciences, DOI 10.1007/978-3-642-16330-2_21,

21.1 Introduction

To be able to develop a territory it is necessary to make works of diverse types that allow creating conditions for exploiting their resources. From the end of the 1960s and later on in the 1980s, when the development in tourism began in this country, the peninsula began to be studied with the objective of knowing its potentials for tourism, verifying the immense possibilities that it offers for the development of a nature tourism for the variety of flora and fauna resources in its emerged portion and submarine platform and the speleologic, archaeological and landscape values.

The peninsula of Guanahacabibes has been little studied due to the inaccessibility that offers its karst relief, the shortage of lands with agricultural potential, the absence of interests for mining activity and geographical accidents that allow the construction of important ports for marine sailing.

Another factor of first importance that breaks the socioeconomic development from the peninsula to a great scale is the enormous deficit of water with quality for its different uses (Vengrechanovich 1980). For its form and geologic constitution, with prevalence in the surface of carbonated karstified rocks, the hydric superficial resources of the peninsula are null and the shallow underground waters are generally affected by marine intrusion processes that convert them into not being capable for their social use, which constitutes the fundamental problem.

Due to the inconveniences of the analyzed variants in Peninsula Guanahacabibes with the intention of another variant starting from a detailed regional conceptual model of the geologic, geophysics, hydrological, hydrogeological and perforation information accumulated until the present in the region, the existence of a regional flow of waters of deep circulation of the Jurassic aquifer constituted by carbonated rocks (calcareous) of the Guasasa Formation of the age $J_3 - K_1$ is possible.

21.2 Geographical Characteristics of the Region

The implied area in the present investigation embraces the western south plain located to the southwest of Guane town as well as the hydrographic basin of Cuyaguateje. The interaction of the geologic, geomorphologic, climatic, hydrograghic and biotic factors have given the place a very specific landscape of lakes, lagoons, swamps, etc.

The peninsula of Guanahacabibes is geomorphologically an active karst plain, where several characteristic types of relief are appreciated, either emerged or submerged (ACC 1989).

To the north of the area occupied by Cuyaguateje Basin there are low mountains, which are very irregular. The average absolute mark of these elevations can be considered between 350 and 400 m; they are of the horst type and blocks of overthrust mantels, especially in the mogotes.

21.2.1 Regional Hydrogeologic Frame

The investigation that was carried out implies the analysis of the characteristics and the role played by three fundamental hydrogeologic units developed in the implied territory that extends from the eastern end of the basin of Cuyaguateje River to the Peninsula of Guanahacabibes. The search of a solution for the water supply in this last unit is not feasible without an integral and interrelated analysis of all the factors that can determine the possible election of variants (Castaño 2000).

- Neogene-Quaternary Complex (N1-Q III): This aquifer is broadly extended for the whole territory. The aquifer is related to the carbonated deposits of the Paso Real, Guane, Vedado and Jaimanitas Formations (Denis and Díaz Guanche 1993) with reef calcareous and organogene strongly karstified in the upper part of the geologic cut to approximately 300 m of depth. In the northern part of the area the rocks of Paso Real lie directly on the deposits of San Cayetano (Well Guane – 1). To the Paso Real Aquifer are attributed very high hydraulic properties with hydraulic conductivities elevated to 700–1,000 m/day, with specific expanses over 25–30 l/s/m and static levels very close to the surface – 1–4 m. The total thickness of the carbonated rocks of the units of the Neogene-Quaternary can range to 750m (ONRM 1970). The underground waters of this aquifer are fundamentally of the chloride-sodium type, due to the marine intrusion with a quantity of soluble salts that oscillates between 1 and 5 g/l, the chlorides vary from 500 to 800 mg/l, in general. In isolated points of the peninsula can be located lenses of fresh water "floating" on the salty waters of bicarbonate-calcic type with a total of dissolved salts of 0.6–0.8 g/l and chlorides between 90 and 290 mg/l (Alvares and Peláez 1970).
- Upper Jurassic – Inferior Cretácico Aquifer Complex (J_3-K_1gs) guasasa: This unit has its biggest development and blooming area in the intermountain hydrogeologic basin Sierra de los Organos coinciding with the hydrographic basin Cuyaguateje. This complex is related with the gray karstified and cracked limestones of the upper Jurassic that appear in the Sierra de los Organos conforming a relief of karst-denudative low mountains. When one advances toward the occident the complex of carbonated rocks submerges and is recovered by the terrestrial silts of the Paso Real Formation until arriving near to Isabel Rubio town where the carbonated complex does not appear for the geologic formation mentioned before. It is as though this whole structure goes toward the southwest of the Guanahacabibes Peninsula. To the north exists an alignment toward the southwest of big karst springs with flow with fresh water that oscillates between 80 and 350 l/s (Springs Los Acosta – 100 l/s, Teneria – 80 l/s and Los Portales 350 l/s) (Chetoni et al 1997). The presence of these springs and other discharges not mentioned attest to the existence of a component of the underground glide of considerable magnitude. A detailed valuation was made. The waters are bicarbonated-calcic type with a quantity of soluble salts between 0.3 and 0.6 g/l, stable in type and content of salts in the time, according to the monitoring of quality of the territory in more than 30 years in which substantial variations do not

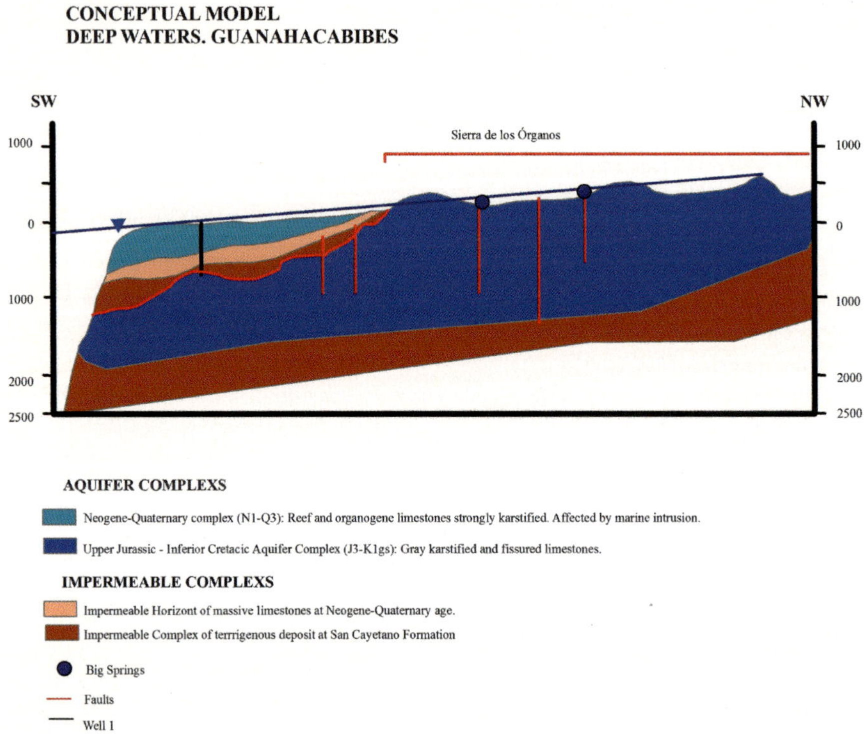

Fig. 21.1 Regional hydrogeologic model

exist in the chemical elements that characterize them. The hydraulic conductivities of this aquifer vary between 50 and 300 m/day. Through this aquifer the underground flow moves toward the southwest, that is, toward the peninsula (Díaz Hernández 2000).

- Upper Jurassic – Inferior Cretacic Aquifer Complex Esperanza (J_3-K_1es): It occupies the north of Cabo San Antonio from the plane of overthrust discovered for the Well Guanahacabibes-1 (Martínez 1994; Martínez et al 1988) that cut sericític carbonaceous schists of black color with andesite intercalations and gabros. The hydraulic conductivities of this aquifer have values up to 100 m/day. This aquifer unit is not considered as perspective (Fig. 21.1).

21.3 Methodology of the Investigation

The new variant is based on a Regional Conceptual Model of the occurrence of a deep aquifer in the Jurassic sediments of Guasasa Formation that underlie the carbonated rocks of the Neogene as well as the terrestrial sediments of San

Cayetano Formation like regional impermeable in the Guanahacabibes Peninsula. This idea is based on the fulfillment of several premises and general conditions that allow a base model of the underground runoff in the territory objective to study the argument of the possibility of capturing abundant flows of underground waters of good quality to depths over 800 m.

The premises that allow the verification of the outlined conceptual pattern are the following:

- Suitable geologic – hydrogeologic constitution.
- Positive hydric balance in the Cuyaguateje basin.
- Existence of submarine discharge areas.

21.3.1 Evaluation of the General Water Balance of Cuyaguateje Basin

The basin of Cuyaguateje River occupies a great part of the western portion of Sierra de los Organos extending from Cabezas town to the outlet of the river in the Ensenada de Cortés, which, in general has a surface of 723 km^2 (CITMA 2005).

Taking into account the environment of general karst development that has a great influence on the hydric balance of the basin, a special interest is dedicated to the valuation of the resources of underground runoff that have their origin inside the limits of the basin, and should go in a southwest direction toward the western is dedicated to part of the Peninsula of Guanacahabibes, taking advantage of the deep continuation of the geologic structure conformed by the carbonated rocks of Guasasa Formation.

The result of the hydric balance in Cuyaguateje Basin settles down for a component of underground runoff toward the south in about 548 million m^3/year (17.3 m^3/s) of water. Starting from the analysis of the structural geologic conditions of the region, this underground runoff should conform to the submarine discharges of the deep Jurassic aquifer in the marine channel along the beril edge to depths over 800 m.

The use of the underground resources that will be discharged to the sea constitutes a potential variant for water supply for the future investments foreseen in the Development Plan of Guanahacabibes that is insignificant of about 30 l/s according to the calculations of these enormous evaluated resources.

21.3.2 Existence of Areas of Submarine Discharges of Water

With the view to predict the places of more probability of submarine discharges a work of localization was carried out of these areas by means of the processing of available satellite images corresponding to the area of the marine platform.

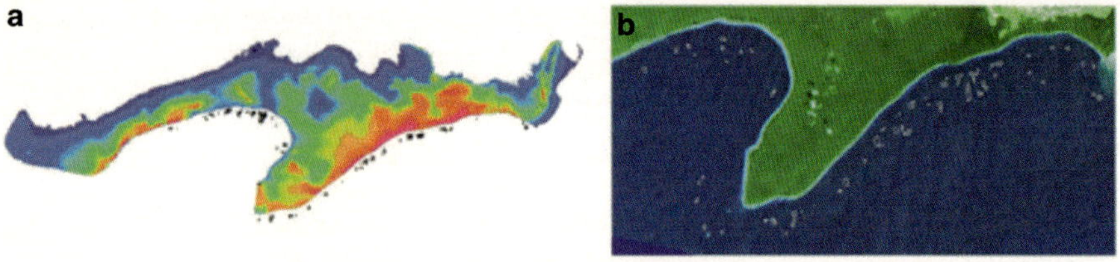

Fig. 21.2 Submarine discharge associated to the aquifer N-Q (*left*). Submarine discharges associated to the deep aquifer of Guasasa Formation (*right*)

In the area to investigate only the information exists related to satellite images (Multispectral reflectance) and bathometry data. The satellite information is limited to a set of images of the Thematic Map (TM) of year 2000 that includes the bands 1, 2, 3, 4, 5, 7, 61 and 62.

The bathometry was extracted from cartographic maps 1:250,000 with an implicit precision.

21.4 Results and Discussions

As a result of the analysis of the regional geologic-hydrogeologic structure the continuity in depth of the sequences of Sierra de los Organos toward the west-southwest in Guanahacabibes region were considered, and can also wait the existence of "valleys and windows" of tectonic-erosive origin under neogene – quaternary carbonated sediments.

The water balance in Cuyaguateje basin throws a component of underground runoff in about 548 million m^3/year (17.3 m^3/s) of water directed toward the south. Starting from the analysis of the structural geologic conditions of the region this underground runoff should conform the submarine discharges of the aquifer detected by the analysis of satellite images in the blooming areas along the insular talus to depths over 1,000 m.

In the Fig. 21.2 (left) the discharges, of the neogene-quaternary aquifer on the isobate of 200–500 m and in the Fig. 21.2 (right) Jurassic aquifer on the isobata of 1,000 m usually model the coast of the peninsula and present exit areas of 500 $\times$ 500 m and larger.

The use of the underground resources that in a certain way will be discharged to the sea constitutes a potential variant for water supply for the future investments foreseen in the Development Plan of Guanahacabibes that is insignificant according to the calculations (30 l/s) in front of these enormous resources.

The Jurassic discharges associated with the Guasasa Formation, according to the Regional Conceptual Hydrogeologic Model, are disseminated between 1,000 and 1,500 m deep. In this interval of depths certain groups appear with a symmetrical correct alignment to the coast. The main direction of calcareous structure toward the peninsula was established by data of the complex analysis with a quite high

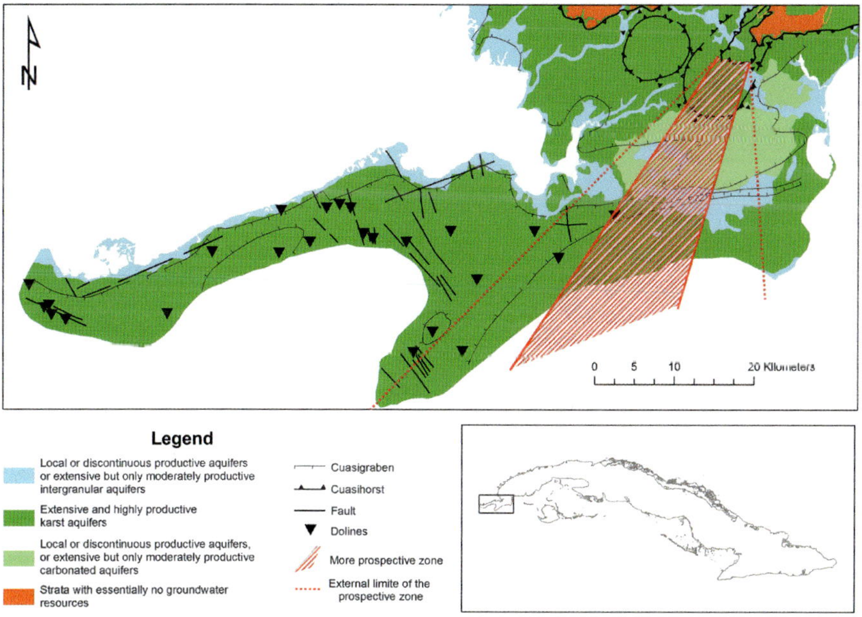

Fig. 21.3 Direction of the prolongation of the deep aquifer structure

security as to have the continuity of the future investigation that allows thinking of a deep perforation of about 1,000–1,200 m of depth in the central area shown in Fig. 21.3.

21.5 Conclusions

According to the analysis of the available information on the geologic structure of the territory, it is possible to assure the occurrence of a deep aquifer container of capable waters for the human consumption, associated with carbonated rocks of Guasasa Formation.

The results of the calculation of the hydric balance in Cuyaguateje River attest for the existence of an underground runoff of large proportions that should circulate for the deep aquifer horizon of Guasasa Formation in the direction of The National Park Guanahacabibes.

Although the available satellite information was not optimum, the investigation showed as a result the existence of submarine discharges of underground waters, located in the area of the blooming of the Guasasa Formation in the marine bottom, between Cabo Frances and Cabo Corrientes to 1,000–1,500 m of depth.

A preliminary evaluation of the economic feasibility of the different variants for water supply in the Peninsula Guanahacabibes allows acceptance of the option of the employment of underground water of deep circulation of the aquifer Guasasa. This will be possible due to its low operation cost and environmental impacts.

In the future geophysical works of electric sounding, gravimetric and seismic to locate fossil valleys covered by Paso Real Formation will be needed to locate a deep exploratory well of 800–1,000 m in depth to carry out hydrogeologic works for the evaluation of the aquifer Guasasa.

References

ACC (1989) Nuevo Atlas Nacional de Cuba. Ed. Instituto de Geografía de la Academia de Ciencias de Cuba

Álvarez Sánchez H, Peláez García R (1970) Informe sobre las investigaciones hidrogeológicas en la Península de Guanahacabibes. Archivo INRH, Inédito

Castaño S (2000) Land and water management in mediterranean islands using earth observation data. Instituto de Técnicas Aeroespaciales de España. http://www.uclm.es

Chetoni R, Reyes R, Martinez D, Sobrino E (1997) Reporte de las investigaciones Geológicas e Hidrogeológicas realizada en los Manantiales “Los Portales”, Guane. Pinar del Río (Archivo ONRM), La Habana, Cuba

CITMA (2005) Plan Gestor de Ordenamiento Territorial de la Península de Guanahacabibes (Archivo Delg. Prov. CITMA). Pinar del Río, Cuba

Denis R, Díaz Guanche C (1993) Características geológicas y geomorfológicas de la Península Guanahacabibes. Dirección Provincial de Planificación Física, Pinar del Río

Díaz Hernández D (2000) Informe sobre los pozos de monitoreo del Centro Internacional de Buceo María La Gorda. Archivo Instituto Nacional de Recursos Hidráulicos (INRH), Inédito

Martínez D (1994) Informe sobre la Generalización del Mapa Geológico del Occidente de Cuba a Escala 1:100000. (Archivo ONRM). La Habana, Cuba

Martínez D, Fernández de Lara R, Peláez R et al (1988) Informe Sobre los Trabajos de Levantamiento Geológico 1:50 000 y Búsquedas Acompañantes. Pinar Central (CAME II), ONRM

ONRM (1970) Reporte sobre el Pozo Guanahacabibes. (Archivo ONRM) La Habana, Cuba

Vengrechanovich R (1980) Informe sobre las investigaciones hidrogeológicas en la Península de Guanahacabibes. Archivo INRH, Inédito

Chapter 22
Aquifer Recharge Capacity in the Protected Area South of the Basin of Mexico

M.F. Naranjo Pérez de León and A. Breña Puyol

Abstract The aim of this study was to conduct a hydrologic analysis to define the boundaries of the 29 watersheds that make up the study area, as well as physiographic characterization through the estimation of parameters governing their hydrological functioning. The results, in summary, is that the evaporation, evapotranspiration and storage in the unsaturated zone, are linked closely to the functioning of the soil where it is concluded that more than two thirds hasty return of the water to the atmosphere through the three processes described. The estimate of recharge capacity or volume of infiltration was carried out through a methodology whose structure is based on three concepts: location, climate, hydrology and hydrographic aspects of micro, water balance of the Valley of Mexico Basin; and water balance of the Conservation Area.

Keywords Watersheds • Aquifer recharge capacity • Hydrology • Mexico City

22.1 Location, Climate, Hydrology and Hydrographic Aspects of Watersheds in the Protected Zones

22.1.1 Location Area

Mexico City is located southwest of the Basin of Mexico; it has a total area of 150,156 ha, 59.04 % of the territory, which corresponds to an area with rural characteristics. In this region, agricultural areas and natural ecosystems play an

M.F. Naranjo Pérez de León (✉)
Researcher Fundación IMDEA AGUA CL, Punto Net s/n, Alcalá de Henares, España
e-mail: naranjo.maria@imdea.org

A. Breña Puyol
Universidad Autónoma Metropolitana, C. del Cubilete No. 302, C.P.04200 México, D.F, Mexico
e-mail: puyol88@yahoo.com

H. Farfán González et al. (eds.), *Management of Water Resources in Protected Areas*, Environmental Earth Sciences, DOI 10.1007/978-3-642-16330-2_22,

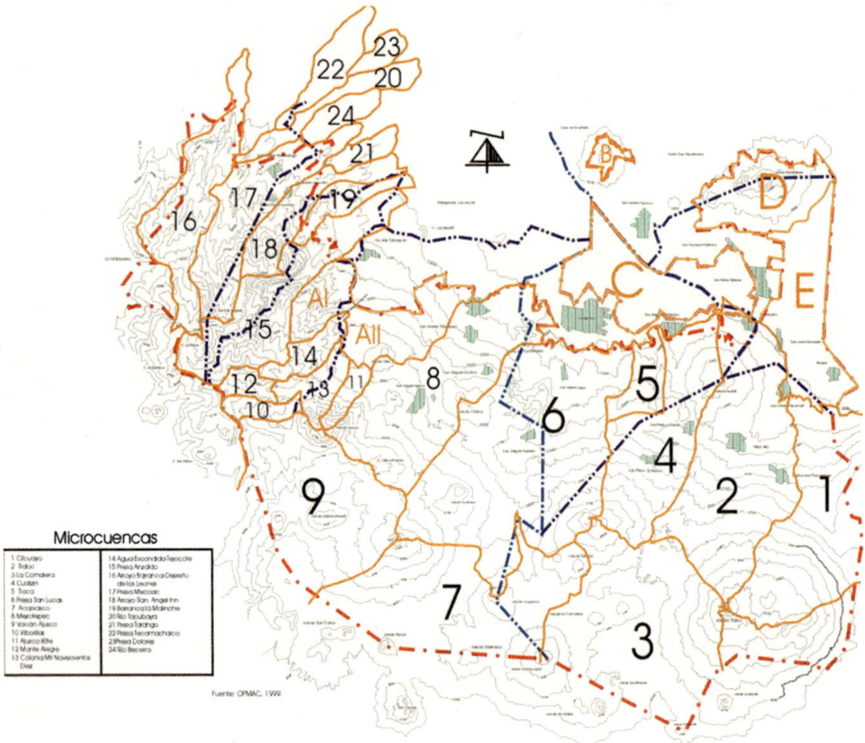

Fig. 22.1 Watersheds

important role in maintaining environmental balance. At an altitude of over 2,240 m.a.s.l, it has four types of climate; and it is characterized by an annual average temperature between 12°C and –18 °C. The rainy season occurs between the months of May to October and the average annual precipitation varies in a range of 600 mm in the northeast, and 1,500 mm in the west.

22.1.2 Hydrology

The aim of the hydrologic analysis conducted on the surface of the Protected Area was to define the boundaries of the 29 hydrologic units or watersheds from the study area.

In each of the hydrologic units a characterization was made through the estimation of parameters such as: area of the basin hypsometric curve (area distribution as a function of elevation); perimeter of the unit, length of the larger axis, elongation ratio, shape index, total length of channels, stream order and drainage density.

The previous parameters define the hydrological behavior of the watersheds, and the results of the analysis are presented in Fig. 22.1:

22.1.3 Hydrographic Aspects of Watersheds

The characteristics presented by the micro-basins are fundamental to understand the hydrological functioning. Figure 22.1 shows the delineation of the 29 watersheds: Basin water balance of the Valley of Mexico, Conceptual Model, Estimation of hydrological balance variables.

22.2 Conceptual Model

The hypothesis used for the water balance is the principle of conservation of mass within the limits established by its watershed. To estimate the water balance of a catchment and to identify the variables involved in this process, the principle of continuity to a control volume that surrounds the main stream was applied, including the eight hydrological variables involved in the balance (Fig. 22.2).

22.2.1 Estimated Balance Hydrological Variables

With the support of the mean values of hydrological variables recorded during a period of 45 years (1960–2004), the water balance of the Basin of Mexico was estimated. Table 22.1 shows the annual average values of its components in hm^3 (million cubic meters) and m^3/s (cubic meters/second) and its corresponding percentages (CNA 2007).

Table 22.2 shows the situation of surface water and groundwater. Analyzing the values in Table 22.2 reveals a number of important conclusions related to the use of surface water and groundwater. In the Mexico City Basin, rainfall runoff generated hm^3 748 (23.7 m^3/s), however, only advantage through rivers and springs a cost of 92 hm^3 (2.9 m^3/s), i.e. only total runoff uses 12 %.

22.3 Water Balance Study, Conceptual Model, Estimation of Hydrological Balance Variables

22.3.1 Conceptual Model

The conceptual model is based on the hydrological processes that occur in the atmosphere, as well as surface and subsurface layers.

In this case, the movements of water in all three phases have a very close relationship, highlighting their behavior in surface and underground layers, as it

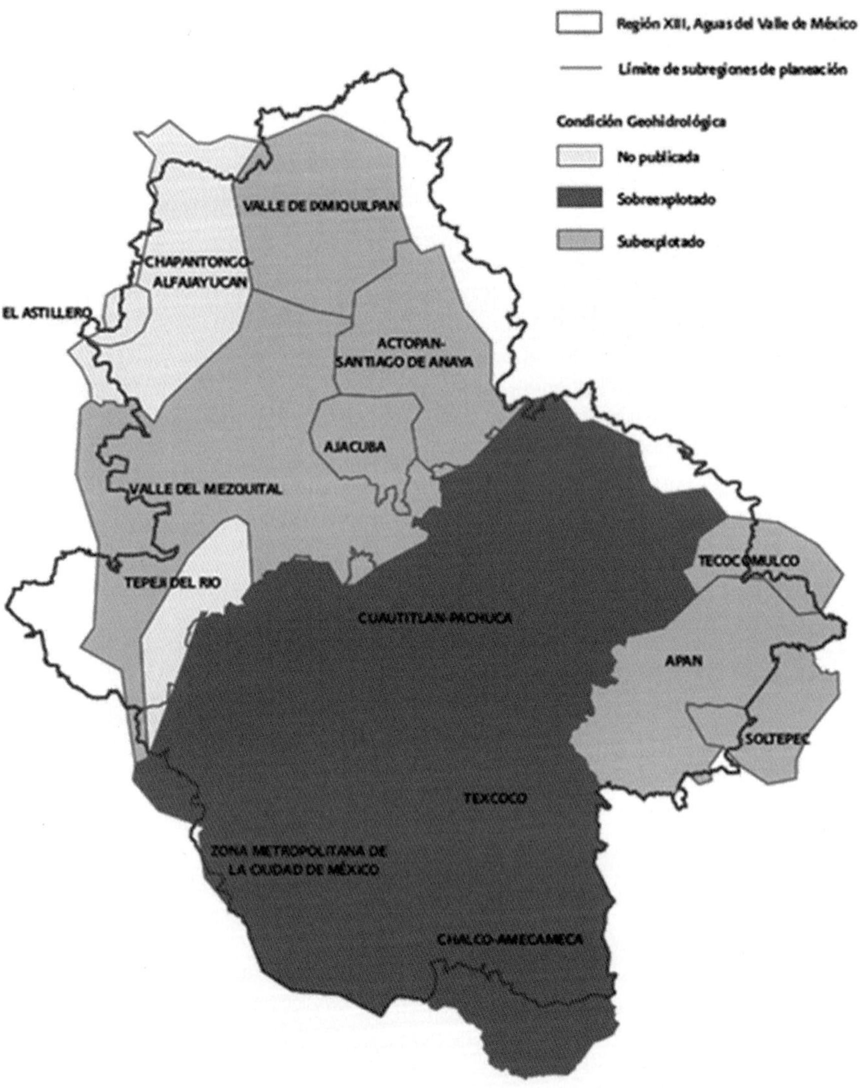

Fig. 22.2 Shows the geohydrological condition of aquifers, overexploited and underexploited

Table 22.1 Hydrological balance for the Basin of Mexico

Component	Volume (hm^3)	Streamflow (m^3/s)	Percentage (%)
Precipitation	6771	214.7	100.0
Evapotranspiration	5014	159.0	74.1
Water storage evaporation	126	4.0	1.9
Superficial recharge	883	28.0	13.0
Runoff	748	23.7	11.0

Table 22.2 Situation of surface water and groundwater in Basin of Mexico

Aquifer	Recharge (hm^3)	Recharge (%)	Concession water volume (hm^3)	Average availability (hm^3)
México city metropolitan area	328	37.2	1249	−921
Cuautitlán-Pachuca	239	27.0	243	−4
Texcoco	57	6.5	93	−36
Chalco-Amecameca	87	9.9	90	−3
Tecocomulco	33	3.7	0	33
Apan	117	13.2	8	109
Soltepec	22	2.5	18	4
Total	883	100.0	1683	−818

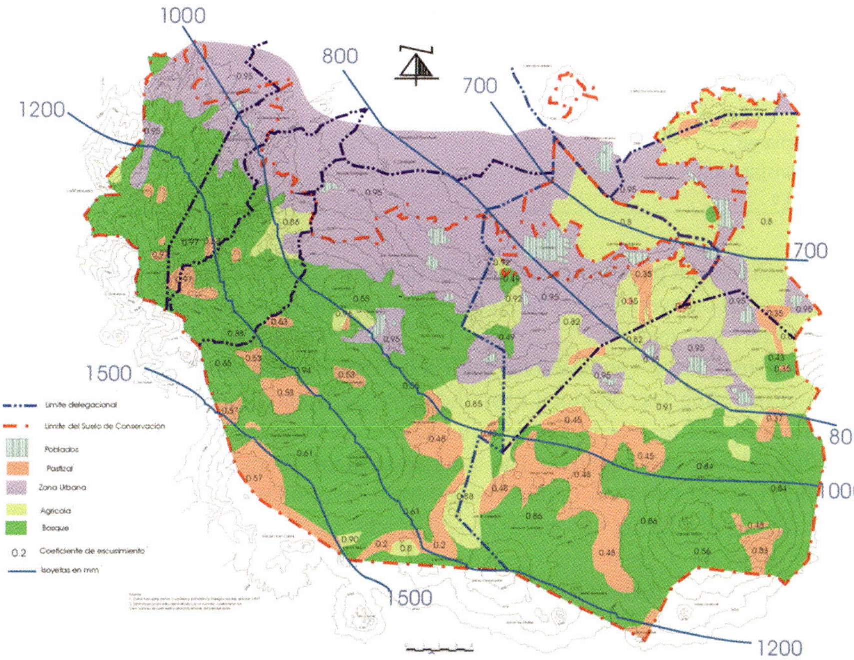

Fig. 22.3 Isohyets in mm OPMAC (1999)

depends on vegetation cover and the dominant subsoil type. Figure 22.3 illustrates the map of isohyets defined, including the current land use and runoff coefficient values. At a later stage, with the support of Fig. 22.3, it upheld the method of isohyets, the height of average rainfall and the areas between two consecutive isohyets. Derived from the values described above, the volume of rain between isohyets was calculated by multiplying the height of average rainfall for the area between isohyets, and adding the partial values, which gives the total volume of rain that rushes over the test area.

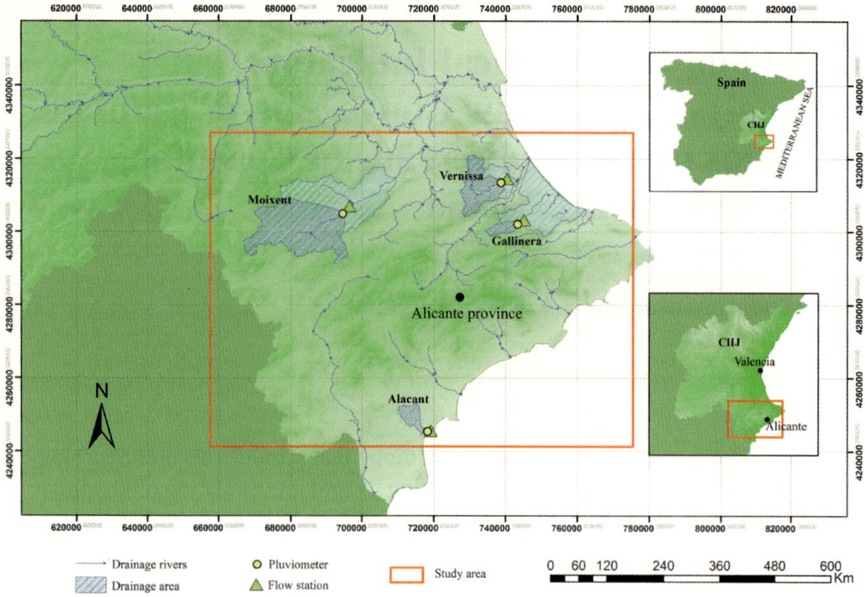

Fig. 23.1 Location of the study area and instrument setting

23.1 Introduction

The study of the relationship between rainfall and runoff has been widely used for the design, forecasting and assessment of water resources in areas sensitive to human interventions, such as protected areas. When there are no runoff data available, or the information is inadequate for an interpretation or reliable extrapolation, rainfall-runoff relationship might be notably useful, as it allows a measurement of runoff from precipitation records, and the hydrological indicators obtained provide highly effective results (WMO 1994; Villón Béjar 2004), improving watershed management as well as the control and efficiency of water resources management.

The aim of this paper is to obtain hydrological indicators to facilitate rapid interpretation of runoff generation from a precipitation event by analysing the conversion of rain in channel flow in four western Mediterranean basins.

23.2 Study Area

The study area comprises the watersheds of Alacant, Gallinera, Moixent and Vernissa (Fig. 23.1), which in turn constitute a series of basins belonging to the Mediterranean coastline, located south of the Alicante province, southeast Iberian Peninsula.

Table 23.1 Key features of the basins analyzed

Basin	P_m* 1991–2009 (mm)	Q_m* 1991–2009 (m^3/s)	Dominant lithology	Total area (km^2)	Drainage area (km^2)	UTM zone 30 X	UTM zone 30 Y
Alacant	20.03	0.172	Pebbles, gravels, sands, silts	31.92	31.92	717,501	4,245,763
Gallinera	80.28	1.201	Conglomerates, marls, sandstones and limestone	224.65	22.63	743,420	4,302,186
Moixent	41.31	0.623	Conglomerates, marls, sandstone and clays	576.49	315.18	694,507	4,305,089
Vernissa	52.70	1.91	Conglomerates, marls, limestone and dolomites	149.45	108.77	738,725	4,313,435

*Pm = Average monthly precipitation
*Qm = Average monthly flow

In terms of hydrogeology, the study area consists mainly of a clump of Caroche Cretaceous, predominantly limestone aquifer, and, to a lesser extent, in the south-eastern coastal area, of a detrital aquifer. Table 23.1 summarises other important characteristics of the watersheds making up the study area.

23.3 Methodology

23.3.1 Data Collection and Preliminary Processing of Information

For real-time data of rainfall and river flows, information from the SAIH (Automatic Hydrological Information System) on the Jucar River Basin supplied by the Ministry of the Environment and Rural and Marine Affairs (MARM) was analysed. Prior to analysis, data were organized to be consistently integrated in the various comparisons drawn. For the systematization of information, different techniques were used in the study, such as: (1) The MARM SAIH Spain, (2) Geographic Information Systems (GIS) to calculate the drainage area (input of runoff) and interpolation of rainfall and (3) the Microsoft Excel program, for the statistical analysis of data, completion of precipitation events and relation of trends between rainfall and water measured in different flow stations.

23.3.2 *Analysis*

Two types of analyses were established as the basis of the research. The first analysis was centered on three indices that make up the rainfall runoff relationship: (a) Depth of direct runoff (Ponce 1997; Cammeraat 2004) or rate of runoff (TE; Pascual et al. 2010), defined as the ratio between the amount of total runoff (Et) of a certain event divided by the surface drainage area (A): TE = Et/A; (b) Coefficient of runoff (C), which relates rate of runoff (TE) and total rainfall after direct runoff begins (ΣRm): C = TE/ΣRm, and (c) Percentage of runoff (PE), which is actually an adaptation of the runoff index, defined as the fraction of total water (Et) of a certain event divided by the surface drainage area: PE = Et/Pt * 100.

The second analysis was based on three comparative hydrological indicators expressing the response of catchments to a rainfall event: (a) The widespread rainfall-runoff relationship (ΦvT) relates the total volume of rain from a precipitation event recorded in a rainfall gauge (Vr) and the total volume of water measured at gauging (Vw): ΦvT = Vr/Vw; (b). The delay time (Tr; Ludwig et al. 2000), which is the length of time between the peak of precipitation (hyetograph) and maximum response as flow (hydrograph), (c) The rate of rainfall intensity (PI), relating the average size of rainfall per square meter (Im); and (d) the equivalent of runoff (Er): PI = Im/Er. All precipitation events were studied, as well as their corresponding production flow based on the use of hyetographs and hydrographs on the watershed drainage area, proceeding to organise the series in chronological order to determine the behaviour of rainfall in each river basin and the response to the event.

23.4 Discussion and Results

The analysis of the precipitation events shows how larger drainage areas are related to a decrease in the runoff rate (TE), runoff coefficient (C) and runoff percentage (PE). Moixent basin, whose drainage area (315.18 km^2, 55 % total area) is the largest, has the lowest runoff production, while Alacant basin, with a drainage area of 31.92 km^2, shows the TE and C. In all basins, runoff production for the event of precipitation from 28 September to 2 October 2009 (Table 23.2) is more representative when compared with the values of rainfall events on watersheds over the 3 years considered (2007, 2008 and 2009). For the event of rainfall recorded, the hydrological response of the four basins is rapid, producing a 1,564 m^3/s average flow of direct runoff, and a delay time of rainfall of less than 2 h ($Tr < 2$ h).

By obtaining hydrological indices, a relationship between rainfalls and the generation of runoff in each basin was noted (Table 23.2). Given the same rainfall conditions, the basins show similar or different behaviours (McKee et al. 1993; Martínez-Mena et al. 1998; Pascual et al. 2010), usually due to the specific features of each drainage area such as their climatic conditions, urban land or land covered by vegetation.

Table 23.2 Results of the hydrological indices

	Alacant			Gallinera			Moixent			Vernissa		
Parameter event	EV1	EV2	EV3	EV1	EV2	EV3	EV1	EV2	EV3	EV1	EV2	EV3
TE ($\times 10^{-2}$ mm)	3.1	0.3	8.8	2.9	0.1	0.23	0.23	0.01	0.5	6.0	1.1	2.0
C ($\times 10^{-3}$)	1.0	1.0	1.0	0.1	0.2	0.1	0.01	0.01	0.01	1.0	0.2	0.2
PE (%)	1.81	1.23	94.5	10.2	50.2	67.3	33.4	14.4	25.6	1.02	4.83	4.69
ΦvT	2.55	6.69	0.85	11.8	10.2	48.4	44.7	56.6	30.1	2.07	67.3	0.18
Tr (hours)	0.50	2.50	3.50	3.25	15.8	1.75	2.50	1.50	1.75	6.25	3.75	4.00
PI	0.07	0.19	0.03	0.47	4.07	1.93	0.13	1.62	0.09	0.02	0.56	0.05

Parameters: TE ($\times 10^{-2}$ mm), Rate of runoff; C ($\times 10^{-3}$), Runoff coefficient; PE (%), Percentage of runoff; ΦvT, Relationship Vr/Vw; Tr (hours), Delay time; PI, Rainfall intensity index
Events: EV1, Event 11–13 Oct. 2007; EV2, Event 09–10 Oct. 2008; EV3, Event 28 Sep–02 Oct. 2009

The highest rate of runoff is recorded in the Alacant basin (8.8×10^{-2} mm), with 3.5 h rain delay. The lowest TE values, close to zero, (0.01×10^{-3} mm) and Tr (1–3 h), occurred in the Moixent basin. Regarding the PE, the highest value (~94.5 %) is recorded in the Alacant basin, and the lowest values, below 10 % (1.02 %), were measured in the Vernissa sector. For each precipitation episode, rain-flow conversion was highly regulated by the drainage area and the inputs, or inputs to the water system (volume, intensity and spatial-temporal distribution of rainfall), thereby characterising the degree of TE, C and PE in the catchments under study. Furthermore, the largest increases in runoff generation in the watershed of Alacant relate both to its drainage area and heavy rainfall in rainy months (September and October) and the high degree of urban land.

In the Gallinera basin, with the smallest drainage area of all watersheds (22.63 km^2, 10 % total area), values are high, medium and low TE, C and PE, which shows that in addition to the drainage area, other factors such as location, climatic characteristics of the basins and rainfall intensity and duration should be taken into account (Yair and Raz-Yassif 2004; Pascual et al. 2010).

The general rainfall-runoff relationship (ΦvT) presents a higher value (67.3) in the Vernissa basin. Regarding the average size ratio per square meter of rainfall and runoff equivalent (assessed by PI, 0–10), similar behavior can be observed in all basins at all events, indicating an intensity precipitation index (PI) of less than unity (PI average 0.77). It should be emphasised that this index is often a good indicator of the magnitude of torrential rainfall (Almorox et al. 2001; Kent and Elliottb 1995). The rainfall volume (Vr) and the volume of water measured at gauging (Vw), calculated to obtain indices, show a narrow relationship ranging from 10 % to 12 % for events that take place in October, and 3–8 % for other events. It is vitally important to note that the systems in the Mediterranean coastal basins have steep slopes and ephemeral movement, which connect coastal and pre-littoral mountain ranges with coastal plains (De Vera 1984; López García et al. 2007). Due to the impact of rainfall and other factors such as urbanization and soil management (Camarasa and Segura 2001; Imeson and Prinsen 2004), these systems underwent major changes in the hydrological response of watersheds analysed when producing runoff.

Chapter 24
The Influences of Precipitations Variability in the Surface Runoff of Hanábana Catchment, Main Contributor of the Largest Wetland in Cuba

K. del Rosario, V. Petrova, and R. L. Winograd

Abstract The purpose of the paper is to show the surface runoff impacts due to variability of rainfall. Hanábana catchment is the main contributor of the Zapata West Wetland, for which a variation in the precipitations would affect the largest Wetland in Cuba directly where the main problem is the water imbalance. The rainfall time series of 78 years (1931–2008) was statistically analyzed to determine the precipitation variability. Three periods were detected in the time evolution of the rainfall, 1931–1960, 1961–1980 and 1981–2008. The first two periods show a growing tendency and the difference between both is not significant. The average of every period corresponds to a probability of means and wet means years, with 36 % and 42 %, respectively. The period 1981–2008 shows a falling tendency of dry characteristics ($p = 63\%$) that could extend up to 2025. The variability of the precipitations to the local scale of the catchment does not coincide with the established periods to global scale for the WMO. The correlation between the runoff and the rainfall demonstrated an average decrease of 10 % of the volume of the contribution toward the Wetland of Zapata.

Keywords Wetlands • Surface runoff • Water imbalance • Hanábana catchment • Cuba

K. del Rosario (✉)
Centro Meteorológico – CITMA, Milanés 27 entre Matanzas y Jovellanos, Matanzas, Cuba
e-mail: katia.rosario@mtz.insmet.cu

V. Petrova
Unidad de Medio Ambiente – CITMA, Milanés 19 entre Matanzas y Jovellanos, Matanzas, Cuba
e-mail: vierapetrova70@yahoo.com

R.L. Winograd
Universidad de Pinar del Río, Martí 270 esq. a 27 de noviembre, Pinar del Río, Cuba

H. Farfán González et al. (eds.), *Management of Water Resources in Protected Areas*, Environmental Earth Sciences, DOI 10.1007/978-3-642-16330-2_24,

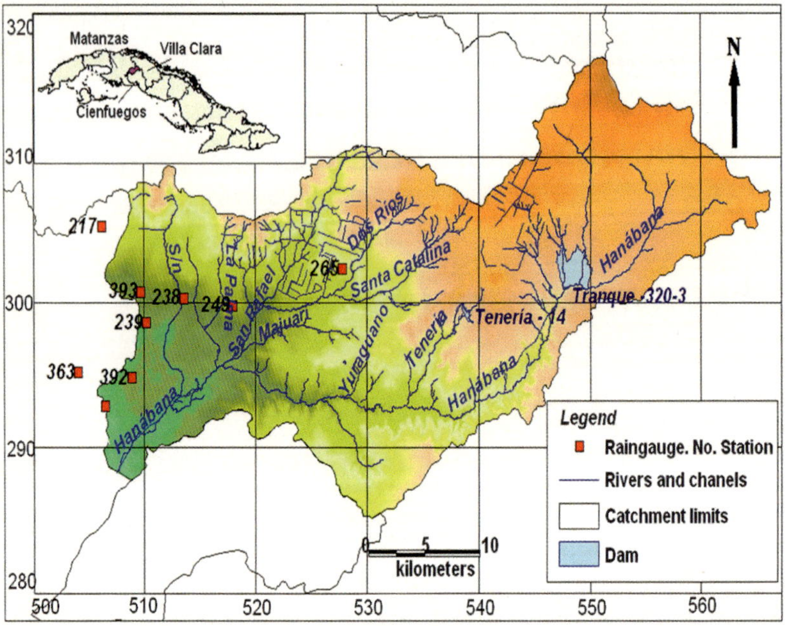

Fig. 24.1 Location of Hanábana catchment

24.1 Introduction

In Cuba, like in the rest of the tropical countries, the main input into the rivers, streams and lakes comes from precipitation, and the readiness of temporary series of rain with more than 50 years allows that the evaluations and predictions that are carried out in a hydrological study have an appropriate physical and logical-mathematical foundation and even more for drainage basins with the short series of hydrometric observations or lack of the same. That is why the purpose of this paper is to establish the influence of the rainfall variability in the surface runoff.

24.2 Study Area

The investigation is developed in Hanábana catchment that is located in the limits of Matanzas, Villa Clara and Cienfuegos Provinces (Fig. 24.1).

This area is presented like a plane plain, with soft slope toward the sea. The heights vary from 80.0 m up to 0.0 m. The basin is in a lengthened way and it narrows, with an area of approximately 1,200 km^2. The longitude of the river Hanábana is 111 km (Petrova 1972) and its largest tributaries are the River San Rafael (Santa Bárbara) and Bagá. This basin is the main contributor of the Zapata West Wetland. The western portion of the Wetland has suffered for more than 40 years countless anthropic modifications. These modifications accompanied by rainfall variability have transformed the water evolution of the Hanábana catchment (Bueno and Petrova 1984).

24.3 Materials and Methods

24.3.1 Rainfall Variability

The rainfall time series of 11 rain gauges were statistically analyzed. The information was obtained from the database of the National Institute of Hydraulic Resources (INRH) in the province of Matanzas. The spatial interpolation and regression analysis was used to correct and supplement the annual data. The presence of trends or inconsistencies referred to as non-homogeneity have been identified by several techniques: double-mass analysis to check the inconsistency of the values of a station with respect to the other neighbors. The trend analysis was carried out by the tests F for the stability of variance and t for stability of the mean; both tests were reinforced by the test of Spearman Rank Correlation. The run test was used to verify the hypothesis of a random order against an alternative of trend.

The variability of rainfall was determined from a single series representative of the basin, spanning from 1931 to 2008, which was studied with the application of several statistical tests as Mann-Kendall and integral differentiated curve. It was also taking into account the periods of 30 years, beginning in 1901, recommended by the World Meteorological Organization in 1994, based on the behavior of the global climate analysis and characterization.

24.3.2 Determination of Runoff

The runoff was computed by three methods.

The analogy between the rivers: Based on geological, hydrogeological, and hydrological studies and climatic conditions of the two watersheds, while taking into account the spatial location (ERP 1989).

Regional Formula for the runoff: Set out in the Regional Scheme (ERP 1989) a regional formula (Eq. 24.1) was proposed for the runoff of the province of Matanzas, which is widely used today for rivers and streams with a very short series of observations or total lack of data.

$$M_0 = 5.42 + 0.055H \tag{24.1}$$

Where:
M_0 – Annual runoff module, expressed in l/seg.Km^2
H – Average height of the basin, in m.
JL Batista's formula (Eq. 24.2), based on rainfall and the average height of the catchment is:

$$R = f_0(0.714\ V_1\bar{P}_1 + 0.244\ V_2H) \tag{24.2}$$

Where:
R – Annual runoff, in mm.
$\bar{P}_1$ – Rainfall, in mm.
H – Average height of the catchment, in m.

24.3.3 Analysis of the Influences of Precipitations Variability in the Surface Runoff

This analysis was based on the relationship between surface runoff and rainfall variability, which was estimated from the simple linear regression method and taking into account the comparison between the periods 1931–1990 and 1991–2008 due to the lack of information and the previous studies.

The calculations of the annual rainfalls and the average of precipitations for 1931–1990 were carried out with eight rain gauges, distributed within Matanzas and Cienfuegos Province (Petrova 1991). For the rainfall calculation of the period 1991–2008, it was not possible to have the eight mentioned rain gauges because some had been disabled. For such a reason, the single series representative of the basin was used, spanning from 1931 to 2008, obtained with the 11 existent rain gauges as was mentioned before.

24.4 Results

24.4.1 Analysis of Rainfall Variability

The spatial homogeneity analysis determined a maximum distance between the gauges equal to 11.8 km, which indicates that it cannot be used for the completion and/or restitution of any of the information that a rain gauge is at a greater distance. The statistical significance of the stability of the variances and the means showed that the 11 rain gauges could be used to calculate a single number representative for the catchment. The calculated representative series covers a period of 78 years (1931–2008) and descriptive statistical parameters are listed in Table 24.1.

The test of Mann-Kendall applied to the annual precipitations series of Hanábana catchment in the period 1931–2008 (Fig. 24.2a), indicated that the series is not homogeneous and it contains tendencies. A significant point of change can be observed among the years 1961 and 1980.

Using the integral differentiated curve of the precipitations (Fig. 24.2b) long-term cycles of different grades of aqueousness have been distinguished: group of mainly humid years (ascending branch) follow for a grouping of dry years (descending branch), routine that repeats in a continuous way. Three fundamental periods are observed, from 1931 to 1960 and from 1961 to 1980, both periods show a growing tendency, while from 1981 until the present time a descending tendency is clearly observed.

Table 24.1 Descriptive statistics of rainfall Hanábana catchment in the period 1931–2008

Characteristic value	Rainfall (mm)
Average	1,472
Absolute minimum	980.5
Minimum characteristic (P 95%)	1,017
Median (P 50%)	1,473
Maximum characteristic (P 5%)	1,852
Absolute maximum	1,950
Standard deviation	238.6
Variation coefficient	16.21

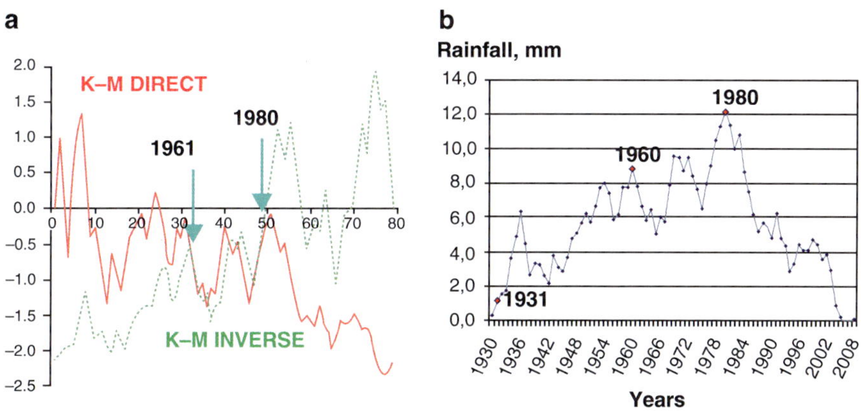

Fig. 24.2 (**a**, **b**) The behavior of Mann-Kendall statistical parameter (direct and inverse, Annual rainfall of the series 1931–2008). (**b**) Integral differentiated curve

The analyses of annual rainfall established the tendency change in 1980. The period begun in this year is defined like the beginning of an average dry period, as it corresponds to 63 % of probability.

The tendency change among the periods 1931–1960 and 1961–1980 are not so evident and the mean value of precipitation in both corresponds to a probability of 36 % and of 42 %, respectively. These values are considered inside the limits of the average and humid average years.

The obtained results are not coincident with the periods settled down by WMO to global scale starting from 1960, that stable 1961–1990 and 1991-present time.

Table 24.2 Different values of runoff in Hanábana catchment by different authors

1959	(NEDECO 1959)	540×10^6 m^3
1973 at 1986	(Petrova 1972; Bueno and Petrova 1984)	300×10^6 m^3
1989	(ERP 1989)	208×10^6 m^3

24.4.2 *Runoff Analysis*

24.4.2.1 Determination of the Surface Runoff Until 1990

The Hanábana Catchment has been studied by different authors in different times obtaining different values of runoff (W_0) as a result (Table 24.2).

In1989 the hydrometric data of the station Hanábana was used and the Station Rodas of Damují watershed was settled down as similar. However, the analysis carried out of the daily and monthly river flow averages showed that in the dry period the flow of the Damují River is greater than Hanábana in 5–10 times and the geologic conditions confirm that the bed of Damují River is deeper and it receives a larger underground contribution than Hanábana. Therefore, it is determined that Damují River cannot be considered as similar to Hanábana River.

24.4.2.2 Runoff Calculation for the Period 1991–2008

For the surface runoff reevaluation of Hanábana Catchment (Petrova 1991), the value of the annual rainfall average computed for the period of 1931–1990 was 1,531 mm. The obtained parameters show that 5 years of hydrometric observations represent very different values of aqueousness (Table 24.3).

The relationship between the surface runoff and the precipitations was determined by the method of lineal simple regression (Fig. 24.3) whose result showed a significant adjustment with a coefficient of determination of 0.97. The model equation obtained was:

$$\text{Esc} = 0.017\ \text{LL} - 16.82 \tag{24.3}$$

Substituting in this equation the mean value of precipitations settled down until that date, 1,531 mm, it was determined that the mean value of runoff of Hanábana River for an area of 952 Km^2, is $Q_0 = 9.20\ \text{m}^3/\text{s}$ and $W_0 = 290 \times 10^6 \text{m}^3$.

Applying the equation of J. Batista (24.2) the values of the mean runoff obtained are: $Q_0 = 8.73\ \text{m}^3/\text{s}$ and $W_0 = 275 \times 10^6 \text{m}^3$.

The volume registered in maps and graphics of the Sciences Academy of Cuba is $W_0 = 309 \times 10^6 \text{m}^3$.

Analyzing the values obtained by the different methods, the surface runoff is in the order of $300 \times 10^6 \text{m}^3$.

Table 24.3 Parameters computed for 5 years of hydrometric observations

Parameters	1964	1965	1966	1967	1968
Rainfall average	1,210	1,028	1,522	1,309	2,105
Discharge (Q)	2.77	1.25	10.8	4.48	18.6
P% average	71.4	90.4	39.8	64.9	4.2

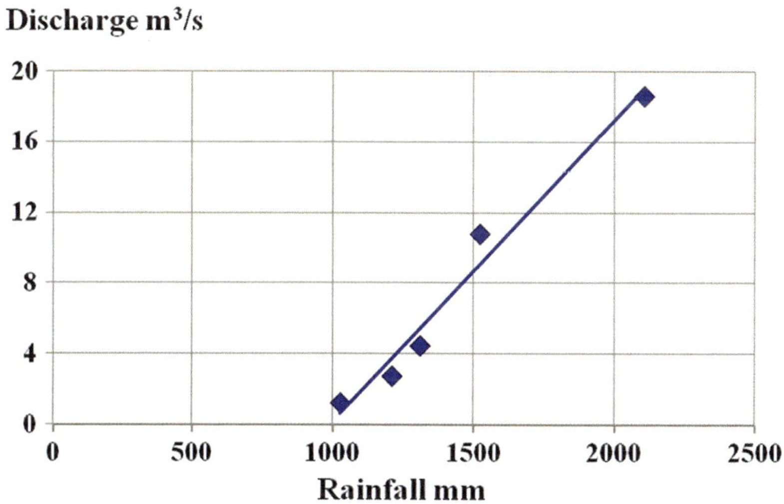

Fig. 24.3 Linear regression line of river discharge against rainfall superposed on scatterplot (Petrova 1991)

24.4.3 *Relationship Between the Surface Runoff and Precipitations*

Because current measurements of runoff do not exist and one of the objectives of this work is to determine the influence of the precipitations variability in the surface runoff, the mean runoff for the period 1991–2008 was computed applying the regression Eq. 24.3. The analysis carried out settles down like mean values of the precipitations, 1,507 mm for 1931–1990 and 1,442 mm for 1991–2008. The runoff for every period was computed substituting the mean rainfall values in the Eq. 24.3. To make some comparison with the previous study (Petrova 1991), it a rainfall value was estimated considering the weight of the eight mentioned rain gauges (Table 24.4).

Keeping in mind that for the runoff data with which it has been carried out these analyses are scarce; and the impossibility of having the same rain gauges with which the first studies were carried out, concludes that there is an affectation between 5 % and 15 % of the surface runoff of Hanábana Catchment due to the rainfall variability.

Table 24.4 Runoff values for every period

Periods	Rain gauges	Rainfall (mm)	Q_0 (m^3/s)	W_0 (10^6 m^3)	%
1931–1990	8	1,531	9.2	290	5
1991–2008		1,499	8.7	274	
1931–1990	11	1,507	8.8	277	15
1991–2008		1,442	7.5	236	

24.5 Conclusions

The surface runoff has fallen between 5 % and 15 % due to the decrease of the precipitations in the period 1991–2008. Considering the predictions that the period of descent of the precipitations will extend up to 2025, it is probable that in the next years a superior decrease of 10 % of the surface contribution toward of Zapata Wetland will occur.

The periods that reflect the changes from the precipitation variabilities to local scale do not coincide, starting from 1961, with the established ones to global scale for WMO. This indicates the necessity to investigate the local precipitation variabilities before proceeding to evaluate and/or to predict a hydrological variable, allowing the making of decisions with the physical and logical-mathematical appropriate foundation.

To guarantee the sustainable development of the territory of Zapata Wetland should be the end of the Program of Water Resources Management (Petrova 2009), to promote the culture of the water care and to reestablish the hydrometric measurements in Hanábana Catchment, which ceased in 1968.

References

Bueno M, Petrova V (1984) Balance hídrico de la Ciénaga de Zapata. Archivos de EIPI – INRH, Matanzas, Cuba, 46p

Esquema Regional Precisado (ERP) (1989) Hidrología de la Ciénaga de Zapata. Archivos del INRH, Cuba

NEDECO (1959) Reclamación de la Ciénaga de Zapata. Habana-Ciénaga. Archivos de la EIPI – INRH, Cuba

Petrova V (1972) Estudio Hidrológico Integral de la cuenca Hanábana. Informe inédito, Archivos del INRH, Cuba

Petrova V (1991) Reevaluación de escurrimiento superficial de la Cuenca Hanábana. Archivos del INRH, Matanzas, Cuba

Petrova V (2009) Programa de manejo de recursos hídricos en los humedales. Caso de estudio: Ciénaga de Zapata. Tesis doctoral, Universidad de Pinar del Río

Chapter 25
Geochemistry of Waters from Tropical Karst Mountain of Western Cuba

J.M. Pajón and J.R. Fagundo

Abstract The studies on geochemistry of the karstic waters, especially, those related to the classic system of chemical reactions CO_2-H_2O-$CaCO_3$, have been fully developed in different karstic areas around the world, many of them to disembowel the mechanism and processes of the karstification. Some significant contributions, among others, developed in the tropical mountain karst of western Cuba are: studies about the chemico-physical behavior of the karstic waters of different hydrogeological natures and their typologies; studies about the chemical evolution of the karstic waters and the empiric relations among variables, parameters and physico-chemical indexes; quantitative estimations about the chemical denudation; the "prompt" work of hydrogeologic, chemico-physical and isotopic monitoring of the karstic waters in different hydrodynamic zones of the karst. This work offers information and brief analyses about the above mentioned, and pay attention to some cardinal problems, totally or partially not yet solved.

Keywords Karst waters • Geochemistry • Isotopic monitoring • Cuba

25.1 Introduction

The researches on geochemistry of the karstic waters, especially those related to the classical system of chemical reactions CO_2-H_2O-$CaCO_3$ and the governing processes and mechanisms of karstification, have been adequately studied (Garrels and

J.M. Pajón (✉)
Museo Nacional de Historia Natural. Obispo 61, Plaza de Armas, La Habana Vieja, La Habana, Cuba, CP: 10 100
e-mail: jesusmpajon@gmail.com

J.R. Fagundo
Chemistry Faculty, Havana University Zapata between G and Carlitos Aguirre, Vedado, PC 10400, Havana
e-mail: juan_fagundo@fq.uh.cu

H. Farfán González et al. (eds.), *Management of Water Resources in Protected Areas*, Environmental Earth Sciences, DOI 10.1007/978-3-642-16330-2_25,

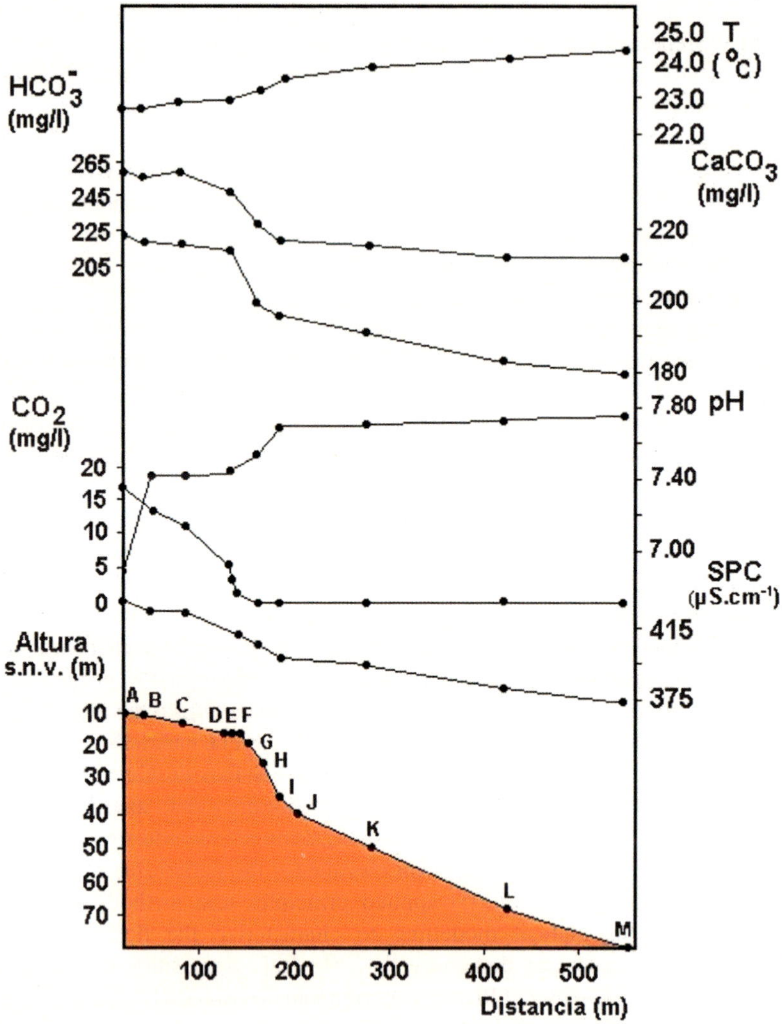

Fig. 25.1 Variation of some parameters and physico-chemical indices of the karstic waters of the Mil Cumbres stream (Sierra del Rosario, Pinar del Río, Cuba). Sampling of the profile realized the 20.09.1985 (Pajón et al. 1990)

as the hardness (expressed as, mg/l de $CaCO_3$), are not linear. To cite a few examples, this behavior is observed in the laboratory work carried out by Picknett (1964), Fagundo and Pajón (1985) and González (1997) in saturated conditions. Fagundo (1985) found that if relationships were determined between ion concentrations and electrical conductivity of waters (SPC) at 25 °C, it is possible to automate the quality control of waters (macrocomponent) by simple measurements of in situ temperature, pH and electrical conductivity. These principles were applied in the karst waters of Pan de Guajaibón with satisfactory results (Fagundo and Pajón 1985).

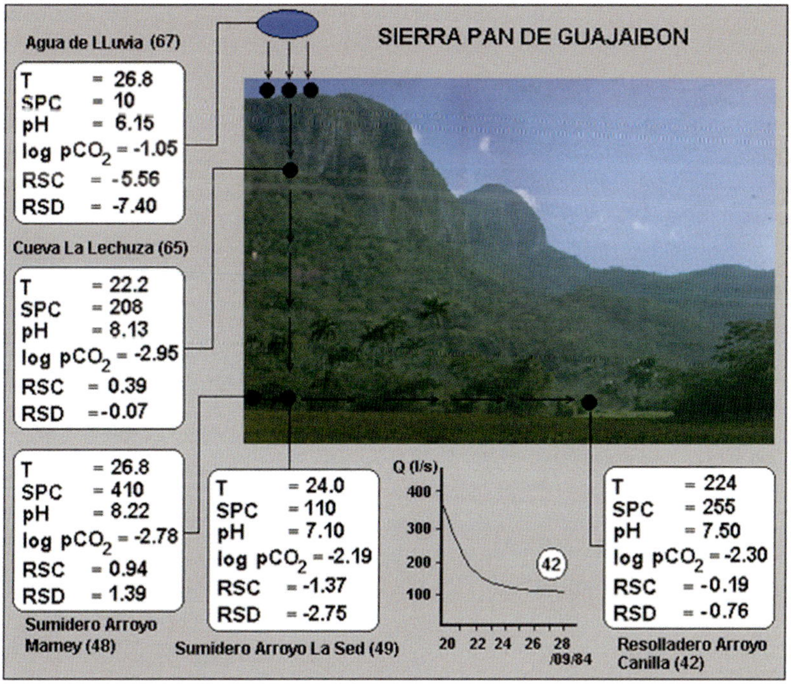

Fig. 25.2 Representative diagram of the model dissolution (spot) Guajaibón karst massif (Campaign 2 September 1984) (Pajón 1990)

Pajón et al. (1990) evaluated as a first approximation, the kinetics of the precipitation of carbonates in the form of travertine along 600 m of the Mil Cumbres stream (Sierra del Rosario, Pinar del Rio, Cuba). They studied pH variation, SPC (25°C), the concentrations of major ionic species macrocomponent, the CO_2 and $CaCO_3$ as well as some chemical and physical indices, to experience the waters from the upwelling to the level of the valley (Fig. 25.1). Pajón (1986) presents a representative diagram of the model dissolution (spot) Guajaibón karst massif, based on the seasonal variations of parameters and physico-chemical indices of waters associated with the main entrances, inside and out sectors of the massif, which represent the areas of food, intermittent aeration and saturation (Fig. 25.2). Fagundo et al. (1991) studied using mathematical modelling, chemical evolution and the empirical relations existing in natural waters of the San Marcos River watershed, Sierra del Rosario, Pinar del Rio, Cuba.

25.2.2 Chemical Denudation

A comprehensive assessment of chemical denudation processes for the seasonally humid tropical karst Pan de Guajaibón can be seen in Pulina et al. (1984),

Fagundo et al. (1987) and Rodriguez (1995). The karst denudation hyperannual values for wet and dry periods, in the Ancon and Canilla karstic systems are as follows:

Ancon System: an average value was obtained for the hydrological cycle of 107 m^3/km^2. year, with 37 m^3/km^2.181 days for dry period and 70 m^3/km^2.184 days for the rainy season. The average values of the order of the highest recorded globally, as in the case of: Gnong Mulu, Indonesia, with 120–200 m^3/km^2.year (Tropical Climate Zone); SW Caucasus, Georgia, 114–139 m^3/km^2.year (Mediterranean climatic zone); Picos de Europa, Spain, 148 m^3/km^2.year (Temperate-Transitional Climatic Zone).

Canilla System: The estimated average chemical denudation for Canilla system was 49 m^3/km^2. year, with 19 m^3/km^2.181 days for dry period and 37 m^3/km^2.184 days for the rainy season. These results are consistent with those reported for other areas of tropical, Mediterranean, temperate and polar climates (Pulina 1977; Garay and Morell 1989; Rodriguez 1995).

The direct linear correlation between denudation and the rainfall has been extensively studied by various researchers for different climatic areas of the world (Pulina (1971); Garay and Morell 1989). Taking into account the values of palaeoprecipitation (up to 6,000 mm) obtained for the karst region of the Sierra de San Carlos during the Sangamon Interglacial (Pajón et al. 2001), replacing them in the equation for calculating the denudation, estimates of chemistry paleodenudation order of those found today in Papua New Guinea can be obtained, where Maire (1981) gave values of chemical denudation in the order of 270–760 m^3/km^2.year, and where ranging from 5,700 to 12,000 mm rainfall annually. Farfán (2005) makes an evaluation processes considering karstification of polycyclic and polygenic natures conducted by the analysis of the age through karstification geoindicators.

25.3 Some Problems and Approaches to the Study of Karst Processes in the Tropical Karst Mountains of Western Cuba

Molerio and Valdés (1975), in a pioneering study ahead of its time, raised some problems and cardinal prospects for the karstological and geospeleological studies in Cuba. They noted, among other things, the importance of hydrochemical investigations in Cuba. Molerio (1981) also noted earlier some problems related to the hydrogeology of the karst mountains of Cuba in general, and Pinar del Rio in particular. The work carried out over several decades in the karst mountain of western Cuba, some of them mentioned in the text, providing various data and assessments that contribute to the understanding of karst processes in this region.

As was mentioned above, monitoring is needed to continue with physical-chemical, isotopical, hydrogeological and climatic stations throughout the zoning of karst hydrodynamics during hydrological cycles appropriate to characterize not only the basic "steady conditions" of karst systems, but also the conditions attached climate-hydrological events diverse and extreme. A notable example in this regard was Project PIGEK (Pulina et al. 1984). Special attention should be paid to the realization of multiparametric monitoring stations in the aeration zone of the karst (hypodermic flow), especially designed to record rainfall and drip in the speleothems. A remaining problem is to obtain quantitative estimates of Global Denudation (DG) in the karst areas, whether mountain or other terrestrial ecosystems in karst. Although there are good estimates of chemical denudation for karst areas (e.g. Pan de Guajaibón), virtually no estimates of mechanical denudation, therefore, has not been possible to have even an estimate of the DG. Moreover, although some efforts have been made, it is important to study the chemical denudation (quantitative estimates) in the coastal karst connected to the karst mountain, taking into account the effect of saline and mixing water on the balance of the carbonate. Of current and capital importance is the continuation of studies related to paleoclimatic and paleoenvironmental reconstructions during the Late Pleistocene-Holocene, from natural paleorecords high resolution, as is the case of speleothems.

Acknowledgments The authors are grateful to Prof. Dr. Augusto Mangini for comment concerning physico-chemistry and isotopic studies and Prof. Luka Dalla Bona who helped revise an earlier version of this paper.

References

Arellano AM, Molerio FL, Suri HA (1992) Efecto de la actitud del ^{18}O en zona de articulación de llanura criptocársica con carso de montaña. In: Llanos HJ, Antigüedad I, Morell I, Eraso A (eds) Taller Internacional sobre Cuencas Experimentales en el Karst. Libro de Comunicaciones, Matanzas, pp 29–42

Back W, Cherry RN, Hanshaw BB (1966) Chemical equilibrion between the water and minerals of a carbonate aquifer. Nat Speleol Soc Bull 28(3):119–126

Bakalowicz M (1979) Contribution de la Geochimie des eaux a la connaissance de L'aquifere karstique et de la karstification. These Doct. es sciences, Paris, 269pp

Boegli A (1972) La Corrosione per miscela d'acque. Atti del Seminario di Speleogenesi. Vorenna. Le Grotte d'Italia IV(4):333–352, 5–8 Otobre 1972

de la Cruz AV, Valdés JJ (1985) Estudio de las aguas kársticas de la Sierra del Pan de Guajaibón y sus inmediaciones mediante métodos matemáticos de análisis exploratorio de los datos. Revista Vol. Hidráulica 68. Año XXII 68:25–34

Dreybrodt W (2000) Equilibrium chemistry of karst water in limestone terranes. In: Klimchouk BA, Ford DC, Palmer AN, Dreybrodt W (eds) Speleogenesis evolution of karst aquifers. National Speleological Society, Huntsville, pp 126–135

Fagundo JR (1985) Caracterización de acuíferos mediante relaciones entre contenidos iónicos y parámetros químico-físicos. Revista CENIC Ciencias Químicas 16(2):321–325

Fagundo JR (1986) Caracterización de acuíferos mediante relaciones entre contenidos iónicos y parámetros químico-físicos. Revista CENIC Ciencias Químicas 16(2):321–325

Fagundo JR (1990) Evolución química y relaciones empíricas en aguas naturales. I- Estudio mediante simulación química del efecto de la litología. Voluntad Hidráulica 82:28–37

Fagundo JR, Pajón JM (1985) Contribución al estudio de las relaciones lineales entre contenidos iónicos y parámetros químico-físicos. Efecto de la litología. Rev Ingeniería Hidráulica VI(1):12–29, ISPJAE, Ciudad de La Habana

Fagundo JR, Valdés JJ (1975) Estudio químico-físico del comportamiento de las aguas kársticas de la región de San Antonio de los Baños, mediante el uso de modelos matemáticos. Ann Speleol 30(4):643–653

Fagundo JR, Pajón JM, Valdés JJ, y Rodríguez JE (1981a) Comportamiento químico-físico de las aguas kársticas de la cuenca del Río Cuyaguateje. Rev Ingeniería Hidráulica 11(3):251–274, ISPJAE, Ciudad de La Habana
Fagundo JR, Pajón JM, Spasov K (1981b) Estudio Hidrogeoquímico de la Cueva Fuentes y sus Alrededores durante la Expedición Espeleológica Búlgaro-Cubana 1981. Revista Voluntad Hidráulica 75(1986):3–8
Fagundo JR, Valdés JJ, Pajón JM (1981c) Estudio Hidroquímico de las Aguas Kársticas de la Cuenca del Río Cuyaguateje. Libro Coloquio Internacional sobre Hidrología Cársica de la Región del Caribe, UNESCO. C, Habana, Dic. 1982, pp 374–401
Fagundo JR, Pajón JM, Valdés JJ, Rodríguez J, Arellano M (1987) Naturaleza de los procesos de disolución de los minerales calcita, dolomita y yeso en las aguas naturales de la región occidental de Cuba. Revista Voluntad Hidráulica 76:30–42
Fagundo JR, Valdés JJ, Cardoso ME, y de la Cruz A (1986a) Algoritmo para el cálculo de parámetros e índices químico-físicos y geoquímicos en aguas naturales altamente mineralizadas. Revista CENIC Ciencias Químicas 12(1/2):72–76
Fagundo JR, Valdés JJ, Rodríguez JE, Pajón JM, de la Cruz A, García A, Pulina M (1986b) Estudio preliminar sobre el proceso de denudación cársica en el polígono cubano-polaco del Pan de Guajaibón. Voluntad Hidráulica 70/71:11–15
Fagundo JR, Pajón JM, Valdés JJ, Rodríguez J, Arellano M (1987) Naturaleza de los procesos de disolución de los minerales calcita, dolomita y yeso en las aguas naturales de la región occidental de Cuba. Revista Voluntad Hidráulica 76:30–42
Fagundo JR, Rodríguez J, Pajón JM, Franco E, Álvarez E, Vinardell J, Vega J, Benítez G (1991) Evolución Química y Relaciones Empíricas en Aguas Naturales. III. Cuenca del Río San Marcos. Sierra del Rosario, Pinar del Río, Cuba. Revista Voluntad Hidráulica No. 85, Año XXVIII, 1991, pp 8–18
Fagundo JR, Rodríguez JE, Pajón JM, Franco E, Benítez G, Rodríguez AC, Güerón J, y Abelló I (1992) Caracterización hidroquímica de las aguas del Pan de Guajaibón y otras áreas cársicas cercanas a la Sierra del Rosario. In: Llanos HJ, Antigüedad I, Morell I, Eraso A (eds) Libro de Comunicaciones I Taller sobre Cuencas Experimentales en el Karst, Matanzas. Universitat Jaume I, Castelló (España), pp 43–53, ISBN: 84-8021-036-2
Ford DC (1988) Characteristics of dissolution cave systems in carbonate rocks. In: James NP, Choauette PW (eds) Paleokarst. Springer, New York, p 601pp
Garay P, Morell I (1989) Tasas de disolución en regiones kársticas españolas. El Karst en España, Madrid, pp 257–264
Garrels RM, Christ CL (1965) Solutions, minerals and equilibria. Harper and Row, New York, p 437pp
González P (1997) Estudio a escala de laboratorio de los procesos de interacción agua-roca. Tesis en opción al título de Maestro en Ciencias del Agua. CNIC, La Habana
González P, Fagundo JR, Antigüedad I, Hernández R, LLanes AJ, Blanco D, Suarez M, Melian C (2005) Catalogo de aguas naturales de la Sierra del Rosario. Banco de Datos del Centro Nacional de Medicina Natural y Tradicional, 44pp
Klimchouk BA, Ford DC, Palmer AN, Dreybrodt W (2000) Speleogenesis. Evolution of karst aquifers. Published by the NES, USA, p 527pp
Langmuir D (1971) The geochemistry of some carbonate ground waters in Central Pennsylvania. Geochim Cosmochim Acta 35(10):1023–1045
Markowics M, y Pulina M (1979) Ilosciowa Polmikroanaliza Chemiczna Wod Wobszarach Krasu Weglanowego. Universytet Slaski, Katowice, pp 67
Miserez JJ (1973) Geochimie des eaux du karst Jurassien. Contribution physico-chimique a l'etude des alterations. These Doct. és sciences. Neuchatel, 313pp
Molerio LF (1981) Problemas hidrogeológicos del Karst de Montaña de Cuba. Voluntad Hidráulica XVIII(56):37–40, La Habana
Molerio LF (1992) Técnicas Estadísticas y de la Teoría de la Información para el Diseño y control de las redes de Monitoreo de las Aguas Subterráneas en el Carso. 1er Taller Iberoamericano de Informática y Geociencias, Acad. Cienc, Cuba, La Habana, p 32

Molerio LF (1995) Regionalización Hidrogeoquímica de las Aguas Subterráneas en la Sierra de Quemados, Pinar del Río, Cuba. Congreso Internacional LV Aniversario de la Sociedad Espeleológica de Cuba y Primera Reunión Iberoamericana, La Habana, pp 92–93

Molerio LF, Valdés JJ (1975) Problemas y perspectivas de la investigación geoespeleológica en Cuba. Revista Tecnológica, 3/75, XIII, pp 34–37

Maire R (1981) Karst and hydrogeology synthesis of Papua-New Guinea. Spelunca 3:7–8

Pajón JM (1986) Geoquímica de las Aguas Kársticas de la Sierra de los Órganos y el Pan de Guajaibón, Pinar del Río, Cuba. Tesis de Grado de Licenciatura en Química, Universidad de La Habana y Centro Nacional de Investigaciones Científicas, 226pp

Pajón JM, Fagundo JR (1978) Aplicación e instrumentación de las técnicas de análisis químico-físico de campo de Markowicz y Pulina a las condiciones de las aguas naturales del karst de la Sierra de los Órganos. Manuscrito Inédito, 50pp

Pajón JM, Fagundo JR (1983) Optimización de métodos analíticos de campo para el estudio de la interacción de las aguas naturales con el medio geológico. Informe Científico-Técnico. Biblioteca BTJ Nacional, 73pp

Pajón JM, García A, Cruz C, de la Cruz A (1990) Evolución Química de las Aguas Kársticas del Arroyo Mil Cumbres (Área Protegida Mil Cumbres), S. del Rosario, Cuba (I). Libro Programa y Resúmenes, Cong. Int. 50 Aniv. de la SEC, pp 114 (25pp)

Pajón JM, Hernández I, Macle J, y Ortega F (2001) Periods of wet climate in Cuba: evaluation of expression in karst of Sierra de San Carlos. In: Markgraf V (ed) Interhemispheric climate linkages (present and past interhemispheric climate linkages in the Americas and their societal effects). Academic, New York, pp 217–226

Pajón JM, Mangini A, Fensterer C, Schroeder-Ritzrau A, Scholz D, Otero V, García G, Chávez O (2009) Monitoreo químico-físico e isotópico de las aguas de flujo hipodérmico de la cueva Dos Anas (Sierra de San Carlos) y la cueva Antorcha (Sierra de Quemados), Sierra de los Órganos, Pinar del Río, Cuba. Informe Científico-Técnico. ICAN, Universidad de Heidelberg, Alemania, 45pp

Pajón JM, Mangini A, Fensterer C, Schroeder-Ritzrau A, Scholz D (2010) Dripwaters Monitoring in cave systems of Western karstic area of Cuba. A preliminary contribution to understanding isotopic composition of speleothems, karst hydrochemistry and kinetic of carbonate dissolution. Part I. Interdynamik project "Caribclim" – Heidelberg – Havana Project. Informe Científico-Técnico (ICT). Banco de Proyectos del ICAN, 20pp

Plummer NL (2002) Geochemical and isotopic studies of paleowaters and recent recharge in parts of the Floridian aquifer system of Florida and Georgia. In Martin JB et al (eds) Hydrogeology and biology of post-paleozoic carbonate aquifers. Proceedings of the symposium karst frontier, Florida and related environments, Gainesville 6–10 Mar 2002, pp 69–72

Pulina M (1971) Observations of the chemical denudation of some karst areas of Europe and Asia. Studia Geomorph Carpatho-Balcanica 5:79–92

Pulina M (1977) Uwagi o zjawiskach krasowyeh w poludniowej czesci Spitsbergenu. Kras i Speleologia, 1, 27pp

Picknett RG (1964) A study of calcite solution at 10°C. Trans Cave Res Group Great Britain 7(1):40–50

Pulina M, Fagundo JR, Valdés JJ, Rodriguez J, Kosik A, Leszkiewicz J, Glowacki P, Pajón JM, de la Cruz AV, García AE (1984) The dynamic of the contemporary karstic processes in the tropical area of Cuba. Preliminary Report of the field investigations performed by the Expedition Guajaibón'84 in the winter season 1984. Universytet slaski, Sosnowiec, Polonia, 42pp

Rodríguez JE (1995) Hidrología y dinámica del karst tropical de montaña de Cuba (Contribución al conocimiento de los sistemas kársticos a partir del ejemplo del Pan de Guajaibón). Tesis para la opción al grado de Doctor en Ciencias Geográficas, La Habana, 150pp

Rodríguez JE, Fagundo JR (1995) Hydrology and dynamics of tropical karst processes in Cuba. Studia Carsologica (Brno) 4: 99–55; (1992). Newsletters 1992, IGCP-299, Guilin (China), pp 46–51

Rodriguez M (2005) Aplicaciones de la Química Física en la caracterización del sistema de flujos de la Sierra del Rosario, Pinar del Río, Cuba. Maestría en Medio Ambiente y Desarrollo. Facultad de Química. La Habana

Stumm W, Morgan JJ (1981) Aquatic chemistry. An introduction emphasizing chemical equilibria in natural waters. Wiley Interscience, New York, p 583pp

Valdés JJ, Fagundo JR, Pajón JM (1981) Aplicación de métodos numéricos al estudio hidroquímico de la cuenca del Río Cuyaguateje. Rev Ingeniería Hidráulica II(3):275–286, ISPJAE, Ciudad de La Habana

White WB (1977) The role of solution kinetics in the development of karst aquifers. In Tolson JS, Doyle FL (eds) Karst hydrogeology (vol 12). International Association Hydrogeology, Memoir, pp 503–517

White WB (2000) Dissolution of limestone form field observation. In: Klimchouk AB (ed) Speleogenesis. Evolution of karst aquifers. Published by the NES, Huntsville, pp 149–155

Wigley TML (1977) WATSPEC: a computer program for determining the equilibrium speciation of aqueous solutions. Brit Geomorph Res Group Tech Bull 20:48pp

Chapter 26
Aspects of the Thermodynamics and Kinetics of the System CO_2-H_2O-Carbonates in Tropical Karst Mountain of Western Cuba

J.M. Pajón and J.R. Fagundo

Abstract The chemico-physical principles, which include the thermodynamic and kinetics basis, of the classic system of chemical reactions CO_2-H_2O-$CaCO_3$, and achieve a control of the mechanism and processes of the karstification, had been fully studied by the international scientific community, with new precisions and contributions in the last decades. Some significant contributions, among others, developed in the tropical mountain karst of western Cuba are: studies about the chemico-physical behavior and the chemical evolution of the karstic waters of different hydrogeological natures; the works about the experimental chemical simulation of the water-rock interaction processes in the karst; the work of hydrogeologic, chemico-physical and isotopic monitoring of the karstic waters in different hydrodynamic zones of the karst. This work offers information and brief analyses about the above mentioned, and attention is paid to some cardinal problems, totally or partially not yet solved.

Keywords Thermodynamic • Karstification • Hydrogeology • Cuba

26.1 Introduction

The chemical and physical principles, including thermodynamic and kinetic fundamentals of the classical system of chemical reactions CO_2-H_2O-$CaCO_3$, and governing processes and mechanisms of karstification have been adequately studied

J.M. Pajón (✉)
Museo Nacional de Historia Natural, Obispo 61, Plaza de Armas, La Habana Vieja, La Habana, CP: 10 100, Cuba
e-mail: jesusmpajon@gmail.com

J.R. Fagundo
Chemistry Faculty, Havana University, Zapata between G and Carlitos Aguirre, Vedado, PC 10400, Havana
e-mail: juan_fagundo@fq.uh.cu

H. Farfán González et al. (eds.), *Management of Water Resources in Protected Areas*, Environmental Earth Sciences, DOI 10.1007/978-3-642-16330-2_26,

(Garrels and Christ 1965; Back et al. 1966; Langmuir 1971; White 1977; Stumm and Morgan 1981; Boegli 1972; Roques 1964, 1973a, b; Picknett 1964, 1972, 1973; Stchouzkoy-Muxart 1972; Miserez 1973; Fagundo and Valdés 1975; Bakalowicz 1979; Ford 1988). New details and contributions have been developed in recent decades, among others, White (2000), Klimchouk et al. (2000), Dreybrodt (2000) and Plummer (2002).

Some contributions to the thermodynamic and kinetic studies of the system CO_2-H_2O-carbonates in the tropical karst mountains of western Cuba are: studies on the chemico-physical behavior and chemical evolution of waters of differing karst hydrogeology (Fagundo et al. 1981a, 1986, 1991, 1992; Fagundo and Pajón 1985; Fagundo 1986, 1990; Pajón et al. 1990); work on the experimental simulation of the processes of water-rock interaction in carbonate karst (Bermúdez et al. 1987; Pajón and Valdés 1991; Fagundo et al. 1992; González et al. 1997; González 1997; Fagundo and González 2003); work "point" monitoring of hydrogeological, chemical, physical and isotopic karst water in different areas of karst hydrodynamics (Fagundo et al. 1985; Molerio et al. 1985, 1999; Arellano et al. 1992; Molerio 1992; Pajón et al. 2009, 2010); paleoclimatic researches from high-resolution records of stable isotopes of oxygen and carbon and Th-U dating and ^{14}C-AMS on stalagmites (Pajón et al. 2006; Pajón 2007, 2009; Fensterer et al. 2010).

Despite the efforts mentioned above, some basic problems remain unresolved, in whole or in part, such as: continuing to conduct monitoring of isotopic chemical-physical, hydrological and climates suitable for hydrological cycles in loading zones stations across the hydrodynamic karst zones; performing multiparametric monitoring stations of the aeration zone of the karst (hypodermic flow), especially designed for recording rainfall and drip into the study of speleothems: palaeoclimate and palaeoenvironmental studies during the Late Pleistocene-Holocene; continuing the simulation experiments of solution chemistry in laboratory conditions in the karst lithologies under n-parametric control in open, closed and "nearly closed" systems; to study the problem of seawater intrusion in the coastal karst connected to the mountain area, related to human activity and climate change, and chemical denudation taking into account the effect of salt and water mixture on the equilibrium of carbonates; deepen the study of the relationship between the CSR (Ratio of the Calcite Saturation) and ΔpH (Tillman-Trombe Index).

26.2 Result and Discussion

26.2.1 Monitoring of the Chemical, Physical, Hydrogeologic and Isotopic Behaviour of the Karstic Waters. Paleoclimatic Researches

Fagundo et al. (1981), using thermodynamic principles, studied the behavior of the main parameters and physico-chemical indexes in 35 sites representing different types of waters of the Cuyaguateje River Basin in the Sierra de los Organos

Mountains (Pinar del Río Province, Cuba). A series of chemical and physical indices, including the relationship of water saturation with respect to calcite (CSR), dolomite (DSR) and gypsum (YSR) were calculated. The data obtained, which characterized the rainy and dry periods 1978–1989, reflected patterns of sequence variation characteristic of the different types of natural waters that occur in the region. A hydrogcochemical study (Fagundo et al. 1981b) was developed in the Fuentes Cave System (Sierra de Mesa, Sierra de los Órganos) and its surroundings, which showed the occurrence of two mechanisms of water-rock interaction. The first one mechanism include lose of CO_2 and carbonates precipitation in the form of travertine along watercourses. Pajón et al. (1990) evaluated as a first approximation, the kinetics of the precipitation of carbonates in the form of travertine along 600 m of the Mil Cumbres stream (Sierra del Rosario, Pinar del Rio, Cuba).

Arellano et al. (1992) performed an interesting study to clarify aspects of the regional flow dynamics and the origin of recharge in the area of articulation of the criptokarstic plain with karst mountain of the Sierra del Rosario (Pinar del Rio, Cuba), particularly in areas related to the Santa Cruz, San Cristobal and Taco Taco rivers. To do this, they monitored the chemical and isotopic composition and physical parameters of the karstic water springs, caves, wells and rainwater in the region, specifically in the basin and Santa Cruz River canyon.

In recent years, paleoclimatic research has been developed in Cuba that has contributed to the assessment of climatic and environmental changes that occurred during the Late Pleistocene-Holocene, from high-resolution records of stable isotopes of oxygen and carbon and Th-U and ^{14}C-AMS dating, on stalagmite representing tropical karst mountains of western Cuba. Oxygen isotope records ($\delta^{18}O$–$\delta^{16}O$ °/$_{oo}$ PDB) and Carbon ($\delta^{13}C$ y $\delta^{12}C$ °/$_{oo}$ PDB) of the studied stalagmites inside the Dos Anas cave (Sierra de San Carlos, Pinar del Rio, Cuba), indicate the following paleoclimatic behavior for the karst mountain area studied (Pajón et al. 1999, 2001a, b, 2006; Pajón 2007, 2009): the existence of large climate variability since the Last Glacial Maximum (LGM) to the present, with a general warming trend; the occurrence of a temperature difference of 8–10 °C between the Last Glacial Maximum (LGM, 18,000 years ago) and the Present Interglacial (Current), confirming the influence of the extent and magnitude of continental cooling occurred in the glacial period on the ecosystems of the tropical-subtropical, especially over the western half of Cuba; detection of abrupt climate warming occurred in the early Holocene, with a high increase of $\approx$ 6-7 °C and that lasted until about 2,000 years ago.

During the 2008–2009 hydrological years, Pajón et al. (2009) developed a bimonthly monitoring program of the karst water infiltration (hypodermic flow) aeration zone of the karst caves, Dos Anas (Majaguas-Cantera Cave System, Sierra de San Carlos) and Torch (Santo Tomás Cave System, Sierra de Quemados), to study in a preliminary way, the behavior of the main chemical-physical parameters and stable isotopes of oxygen and carbon, climate and trace elements at four sites representing these localities. Fensterer et al. (2010) studied by ^{230}Th/U a Late Holocene stalagmite in Dos Anas Cave, whose findings are part of the research program developed in the karst of the Sierra de San Carlos and Sierra de Quemados, to study, using stable isotopes and isotopic dating, the natural behavior of the past climate of Cuba.

26.2.2 Thermodynamic and Kinetic Aspects in the System CO_2-H_2O-$CaCO_3$

From the thermodynamic principles, taking into account the classic equation that represents the dissolution-precipitation of limestones, and a set of chemical reactions and their equilibrium constants, Fagundo and Valdés (1975) arrived at the following expression (Eq. 26.1) to calculate the Coefficient of the Calcite Saturation (CSC):

$$CSC = \left[Ca^{2+}\right]\left[HCO_3^-\right]10^{pK_c - 2pK_2 + pH} \tag{26.1}$$

Where: $[Ca^{2+}]$: Molal concentration of Ca^{2+}, $[HCO_3^-]$: Molal concentration of HCO_3^-, pK_C: $-\log K_C$ (solubility product of calcite), pK_2: $-\log K_2$ (Equilibrium constant for dissociation of the ion HCO_3^-), pH: $-\log [H^+]$.

As the ratio of water saturation with respect to calcite (CSR) is defined as (Eq. 26.2):

$$CSR = \log CSC \tag{26.2}$$

Equation (26.3) is:

$$CSR = \log\left[Ca^{2+}\right] - \log\left[HCO_3^-\right] + pk_c - pK_2 + pH \tag{26.3}$$

That is only the Saturation Index (SI) of Back et al. (1966), then (Eq. 26.4):

$$CSR = IS \tag{26.4}$$

Roques (1972) and Stchouzkoy-Muxart (1972) proposed equations (Eq. 26.5) relating the field measured pH (pH_m) and equilibrium pH (pH_e), to assess the status of the solution referred to the chemical balance, being defined ΔpH index as:

$$\Delta pH = pH_m - pH_e \tag{26.5}$$

The terms proposed by Stchouzkoy-Muxart (1972) and Roques (1972) are:

$$\Delta pH = pH_m - 2\log\left[Ca^{2+}\right] - p\gamma Ca^{2+} - p\gamma HCO_3^- + pK_2 - pK_c - 0.3013 \tag{26.6}$$

$$\Delta pH = pH_m - \log K_s + \log K_2 + \log\left[HCO_3^-\right] - \log\left(1 - K_5\left[SO_4^{2-}\right]\right) + C \tag{26.7}$$

Where: $p\gamma Ca^{2+}$: $-\log$ the activity coefficient of Ca^{2+}, $p\gamma HCO_3^-$: $-\log$ the activity coefficient of HCO_3^-, pK_C: $-\log K_C$ (solubility product of calcite), pK_2: $-\log K_2$ (Equilibrium constant for dissociation of the ion HCO_3^-), K_S: Equivalent to K_C (solubility product of calcite in equilibrium), K_5: Equilibrium constant for the dissociation of $CaSO_4$, $[SO_4^{2-}]$: molal concentration of SO_4^{2-}, Me: may be $[Ca^{2+}]$ o $[Mg^{2+}]$,

C: Represents the contribution to the pH of the ion pairs $MgHCO_3^+$, $MgCO_3^0$, $CaHCO_3^+$ and $CaCO_3^0$, can be calculated as: (Roques 1972 en Bakalowicz 1979, 1980).

Pajón et al. (1985a) studied the correlation between the values of CSR (IS) and ΔpH (expression of Stchouzkoy-Muxart 1972) in natural waters of the karst mountains of the Sierra de los Organos (geochemical testing cycles of the Cuyaguateje River Basin during the periods of March 1979, August 1979, February 1980 and July 1980) and Pan de Guajaibón (geochemistry testing cycles for the period from January to February 1980). They found a numerical similarity between the two indices, considering or not the complex ions and ion pairs (in the calculation of CSR), although the accuracy is higher for the CSR to present them. Based on the work of Berner and Morse (1974) and through the study of carbonate aquifers waters of Pennsylvania and Kentucky, White (1977) found that the parameter ΔpH of the experiments of these authors is related to the IS of the form following:

$$\Delta pH = \frac{1}{2} IS \tag{26.8}$$

The results obtained by Pajón et al. (1985) and Pajón (1986) are consistent with the statements by Bakalowicz (1980), which concludes that the Back saturation index (SI) is equal to the ratio between the measured pH in a solution and the pH of a solution in equilibrium with a given mineral, regardless of ion-pair complexes eventually formed. Although the same author (Bakalowicz 1979) posed that the ΔpH has the same meaning as the IS Back, even if they do not exactly correspond. He thus concludes that:

$$CSR = IS = \Delta pH \tag{26.9}$$

Carbon dioxide plays a fundamental role in the dissolution of carbonates (Miserez 1973; Roques 1973a; Wigley 1973; Bakalowicz 1979; Kempe 1982). Humic and fulvic acids also play an important role in the processes of water-rock interaction of the system CO_2-H_2O-$CaCO_3$. Stchouzkoy-Muxart (1972) makes a numerical and graphical analysis where the pCO_2 is based on the ion HCO_3^- and Ca^{2+} and the equilibrium constants. Fagundo (1982) presents an empirical expression for calculating the pCO_2, which relates to the $\sqrt{I}$ and activity of ion HCO_3^- with the electrical conductivity and temperature.

$$\log pCO_2 = -\frac{A\sqrt{I}}{1 + a^0 HCO_3^- B\sqrt{I}} + \log\left(HCO_3^-\right) + pK_B + pK_1 - pH \tag{26.10}$$

Where: I: Ionic strength, a^o: Effective diameter of the ionic species I, B: Constant B of the Debye-Huckel equation, pK_1: $-\log K_1$ (Equilibrium constant for the dissociation of H_2CO_3), pK_B: $-\log K_B$

26.2.3 Experimental Simulation and Kinetic Studies of the Reaction of Carbonate Dissolution in Kars

Laboratory experimental work with controlled simulation of the processes of water-rock interaction of the system CO_2-H_2O-$CaCO_3$ developed by several authors (Weyl 1958; Picknett 1964; Curl 1965; Stchouzkoy-Muxart 1972; Roques 1973a; Berner and Morse 1974; Rauch and White 1977; Plummer et al. 1978; Richard and Sjöberg 1983; Sjoberg and Rickard 1984; Herman et al. 1986; Comton and Unwin 1990), were a source of inspiration for the early 1990s, and resulted in the first attempts to develop work of this nature with geological samples of carbonates in the Cuban karst regions. These regions develop significant karstification processes that have resulted in the occurrence of large cave systems in Cuba, as is the case of Santo Tomás Cave System, the Majaguas-Cantera Cave System and Palmarito Cave System, among others. Some excellent studies influenced by this methodology were developed by Fagundo et al. (1992), Fagundo and Gonzalez (1997), González et al. (1997) and González (1997).

Based on laboratory kinetic studies on the natural systems of the dissolution of carbonates, Pajón and Valdés (1991) designed a simple solution simulation experiment with rock samples from the formations Guajaibón (RAC-3) and Chiquita (SA-4) belonging to the mountains of the same name in the Pan de Guajaibón massif (Sierra del Rosario, Pinar del Rio, Cuba), as well as a "Tinajitas" (TIN-1), lapiaz in limestone surface of the Ancón Creek in the Pan de Guajaibón massif. The aim of this study was to compare quantitative experimental results concerning the speed and intensity of dissolution of carbonate rocks with chemical denudation estimates derived from field investigations. The experiments described here are of the "free trend" variety (Rauch and White 1977), where an initial solution of distilled water maintained as "quasi-closed" and in equilibrium with pCO_2, reacts with carbonate rocks to a balanced approach. It was noted the purchase of ions HCO_3^- and Ca^{2+} over time, increases in the order of TIN-1 $\rightarrow$ SA-4 $\rightarrow$ RAC-3, which means an increase in the rate of dissolution in the same order. The velocity of dissolution of the Chiquita formation rocks is higher than Guajaibón formation, and in turn in the Tinajitas.

A suitable and systematic study on the kinetics and thermodynamics of the system CO_2-H_2O-Carbonate was carried out by Fagundo et al. (1992), González et al. (1997) and González (1997), which provided the basis for the work of experimental simulation of dissolution of carbonates developed by these authors (Fig. 26.1). There are various procedures and equations to estimate the rate of dissolution in the system CO_2-H_2O-$CaCO_3$, which represents the variation of the concentrations of dissolved species over time, may include the equations of Plummer and Wigley (1976), Sjöberg (1976), Plummer et al. (1978), Christoffersen and Christoffersen (1979), Rickard and Sjöberg (1983), and Appelo and Postma (1993). Fagundo et al. (1996), Álvarez et al. (1996) and González et al. (1997) developed kinetic expressions of the rate of dissolution of carbonates, especially evaluated for the case of carbonate karst geological formations in western Cuba, which have unique cases of open and closed systems for CO_2 (Fagundo and González 2003). Whereas the dissolution of calcite and dolomite in the work of simulation of the water-rock interaction in the laboratory is obtained a

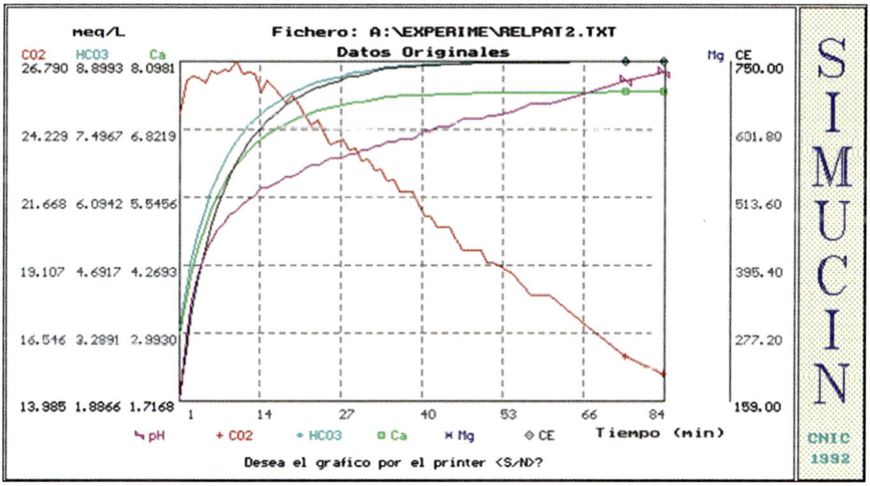

Fig. 26.1 Experimental simulation of dissolution of carbonates (Cinepat 2, limestone). The variation in the time of the pH and the composition of the ions HCO_3^-, Ca^{2+} and Mg^{2+} (González 1997; González et al. 1997; Fagundo and González 1997)

pseudo first-order equation which expresses the change in ionic concentration over time (Fagundo 1996; Álvarez et al. 1996):

$$C = C_{eq}\left(1 - e^{-kt^n}\right) \tag{26.11}$$

Where, n is an experimental coefficient takes values between 0 and 1.

Fagundo et al. (1992) used samples of the limestone formations Jaimanitas, Guasasa, Jagua and Guajaibón in a laboratory experiment, which simulated the process of dissolution of these carbonates under conditions of open and closed system with respect to CO_2. Álvarez et al. (1996) performed simulation experiments in the laboratory using carbonate rocks containing calcite and dolomite minerals in different proportions with the aim of studying the process of acquisition of the chemical composition of natural waters run at a karst aquifer. González (1997) and González et al. (1997) simulated in the laboratory the process of dissolution of carbonate (calcite and dolomite) under conditions of open and closed system for CO_2. They obtained rate expressions that describe the acquisition process of the chemical composition of natural waters that interact with the limestone and dolomite studied. They demonstrated that in closed system conditions during the kinetic process of water-carbonate rock interaction, the variation of CO_2 can be expressed by exponential equations of a specific type, while the change in open system conditions is negligible. The variation in the time of the electrical conductivity and the composition of the ions HCO_3^-, Ca^{2+} and Mg^{2+} can also be expressed with an exponential equation type. The dissolution rate of calcite was found to be higher by about five times compared to dolomite.

26.2.4 Some Problems and Approaches to the Study of Karst Processes in the Tropical Karst Mountains of Western Cuba

Molerio and Valdés (1975), in a pioneering study ahead of its time, raised some problems and cardinal prospects for the karstological and geospeleological studies in Cuba. The authors drew attention to the need to address these studies by weighting the morphogenetic criteria on morphographics and the morphologicals. Here is implied the use of thermodynamic and kinetics approach in the studies on geochemistry of karst waters and paleoclimate from speleothems records (Pajón et al. 2009).

Monitoring is needed to continue with physical, isotopic, chemical, hydrogeological and climatic stations throughout the zoning of karst hydrodynamics during hydrological cycles appropriate to characterize not only the basic "steady conditions" of karst systems, but also the conditions attached climate-hydrological events diverse and extreme. Special attention should be paid to the realization of multiparametric monitoring stations in the aeration zone of the karst (hypodermic flow), especially designed to record rainfall (drip) in the speleothems. A derivation of the monitoring of chemical characteristics and isotopy of seepage water is towards the paleoclimatic and paleoenvironmental studies covering the Late Pleistocene-Holocene.

Despite the unquestionable value of experimental studies of chemical simulation of carbonate dissolution in laboratory conditions from Cuba, it is necessary to continue this line with lithologies, which is heavy in the development of karst. These experiments will lead to a multi-control n-parametric, in open systems, closed or "nearly closed", automated facilities that control a set of variables and parameters previously selected. In this sense the conceptual model developed by Rauch and White (1977) may be resumed, since the experimental design is based on the use of carbonate blocks with ducts simulating karst cavities or channels. The fundamentals developed by Dreybrodt (1992, 2000) can be used, who proposed a model of evolution of karst channels to simulate the dissolution of limestone from these primary cracks.

It remains to explain if a numerical difference exists between the chemical and physical indices of the saturation ratio of calcite (CSR) and Tillman-Trombe Index (ΔpH). The first has two approaches, one by the American school and another school in Cuba, while the ΔpH responds to the European school. There are similarities between the two indices if we consider the Cuban and European schools; however, to compare the rates according to American and European schools is a ½ relationship between them. This is an algorithmic error in this treatment and/or variables that make up the equations, or errors of other kinds?

Of current and capital importance is the continuation of studies related to paleoclimatic and paleoenvironmental reconstructions during the Late Pleistocene-Holocene, from natural paleo records high resolution, as is the case of speleothems.

Acknowledgments The authors are grateful to Prof. Dr. Augusto Mangini for comment concerning physico-chemistry and isotopic studies and Prof. Luka Dalla Bona who helped revise an earlier version of this paper.

References

Álvarez E, Fagundo JR, González P, Ferrera V, Benítez G (1996) Aplicación de métodos de simulación química y matemática al estudio de la disolución de carbonatos en diferentes condiciones experimentales. In: Fagundo Castillo JR, Pérez Franco D, Álvarez Nodarse A, García e I. Morell JM (eds) Contribuciones a la hidrogeología y medio ambiente en Cuba. Universidad de Castellón, España, pp 91–103

Appelo AJ, Postma D (1993) Geochemistry, groundwater and pollution. A.A. Balkema, Rotterdam/Brookfield, p 536pp

Arellano AM, Molerio FL, Suri HA (1992) Efecto de la actitud del ^{18}O en zona de articulación de llanura criptocársica con carso de montaña. Llanos HJ, Antigüedad I, Morell I, Eraso A (eds) I Taller Internacional sobre Cuencas Experimentales en el Karst. Matanzas, Cuba, Abril 1992. Libro de Comunicaciones, pp 29–42

Back W, Cherry RN, Hanshaw BB (1966) Chemical equilibrium between the water and minerals of a carbonate aquifer. Nat Speleol Soc Bull 28(3):119–126

Bakalowicz M (1979) Contribution de la Geochimie des eaux a la connaissance de L'aquifere karstique et de la karstification. These Doct., es Sciences, Paris, 269pp

Bakalowicz M (1980) Un precieux informateur Hydrogeologique: Le systeme chimique CO_2-H_2O-Carbonate. Extrait de: Cristallisation, deformation, dissolution des carbonates. Colloque Soc. Geol. Fr., Assoc. Geol. Sud-Ovest, Soc.Fr. Mineral. Cristall. Bordeaux, 23 pp

Bermúdez T, Fagundo JR, Pajón JM (1987) Cinética de recombinación iónica en aguas cársicas expuestas a la evaporación. Revista CENIC Ciencias Químicas 18(2/3):141–144

Berner RA, Morse JM (1974) Dissolution kinetics of calcium carbonate in sea water: IV theory of calcite dissolution. Am J Sci 274:108–134

Boegli A (1972) La Corrosione per miscela d'acque. Atti del Seminario di Speleogenesi, Vorenna, Le Grotte d'Italia (4) IV, pp 333–352, 5–8 Otobre 1972

Comton RG, Unwin PR (1990) The dissolution of calcite in aqueous solution at $pH < 4$; kinetics and mechanism. Phil Trans R Soc Lond A 330:1–45

Curl RL (1965) Solution kinetic of calcite. In: Proceedings of the 4th international congress, Speleo. Ljubljana, vol 3, pp 61–66

Dreybrodt W (1992) Dynamics of karstification: a model applied to hydraulic structures in karst terraines. Appl Hydrogeol 3(92):20–32

Dreybrodt W (2000) Equilibrium chemistry of karst water in limestone terranes. In: Klimchouk AB et al. (eds) Speleogenesis. Evolution of karst aquifers, January 2000 edition. Published by the NES, pp 126–135

Fagundo JR (1982) Determinación de índices para la caracterización de aguas cársicas mediante formulas semiempíricas. Actas de Primer Coloquio Internacional sobre Hidrología Cársica de la Región del Caribe. Palacio de las Convenciones, La Habana, Cuba, pp 499–509

Fagundo JR (1986) Caracterización de acuíferos mediante relaciones entre contenidos iónicos y parámetros químico-físicos. Revista CENIC Ciencias Químicas 16(2):321–325

Fagundo JR (1990) Evolución química y relaciones empíricas en aguas naturales. I- Estudio mediante simulación química del efecto de la litología. Voluntad Hidráulica 82:28–37

Fagundo JR (1996) Procesos geoquímicos y calidad de las aguas. In: Fagundo Castillo JR, Pérez Franco D, Alvarez Nodarse A, García JM, Morell I (eds) Contribuciones a la hidrogeología y medio ambiente en Cuba. Universidad de Castellón, España, pp 177–198

Fagundo JR, y Pajón JM (1985) Contribución al estudio de las relaciones lineales entre contenidos iónicos y parámetros químico-físicos. Efecto de la litología. Revista Ingeniería Hidráulica VI (1):12–29, ISPJAE, Ciudad de La Habana

Fagundo JR, González P (2003) Estudio indirecto de la carsificación mediante métodos experimentales y mediciones de campo. In: CD Rom Congreso 65 Aniversario de la Sociedad Espeleológica de Cuba, El Moncada, Viñales, Pinar del Río, Cuba, 5pp, 8–12 Febrero. 2003

Fagundo JR, Valdés JJ (1975) Estudio químico-físico del comportamiento de las aguas kársticas de la región de San Antonio de los Baños, mediante el uso de modelos matemáticos. Ann Speleol 30(4):643–653

Fagundo JR, Álvarez E, Benítez G, Ferrera V, y Vega J (1996) Simulación química y matemática de la disolución de rocas carbonatadas por las aguas naturales. Kras i speleologia. T.8, XVII (Katowice):78–90

Fagundo JR, González P, Benítez G, Jiménez S, Santiesteban ER, González A, Romero E (1999) Procesos geoquímicos y calidad de las aguas en los acuíferos costeros. En: El Caribe. Contribución al conocimiento de su Geografía. Ed. Instituto de Geografía Tropical, Agencia de Medio Ambiente, Ministerio de Ciencia, Tecnología y Medio ambiente, pp 40–47

Fagundo JR, Pajón JM, Valdés JJ, Rodríguez JE (1981a) Comportamiento químico-físico de las aguas kársticas de la cuenca del Río Cuyaguateje. Rev Ingeniería Hidráulica, ISPJAE, Ciudad de La Habana 11(3): 251–274

Fagundo JR, Pajón JM, Spasov K (1981b) Estudio Hidrogeoquímico de la Cueva Fuentes y sus Alrededores durante la Expedición Espeleológica Búlgaro-Cubana 1981. Revista Voluntad Hidráulica 75:3–8, 1986

Fagundo JR, Valdés JJ, Cardoso ME, y de la Cruz A (1986) Algoritmo para el cálculo de parámetros e índices químico-físicos y geoquímicos en aguas naturales altamente mineralizadas. Revista CENIC Ciencias Químicas 12(1/2):72–76

Fagundo JR, Rodríguez J, Pajón JM, Franco E, Álvarez E, Vinardell J, Vega J, y Benítez G (1991) Evolución Química y Relaciones Empíricas en Aguas Naturales. III. Cuenca del Río San Marcos. Sierra del Rosario, Pinar del Río, Cuba. Revista Voluntad Hidráulica XXVIII(85):8–18

Fagundo JR, Pajón JM, Valdés JJ, Rodríguez J, Arellano M (1987) Naturaleza de los procesos de disolución de los minerales calcita, dolomita y yeso en las aguas naturales de la región occidental de Cuba. Revista Voluntad Hidráulica 76:30–42

Fagundo JR, Álvarez E, Benítez G, Ferrera V, Vega J (1992) Simulación química y matemática de la disolución de rocas carbonatadas por las aguas naturales. Proceedings of Del XXIII Congreso Interamericano de Ingeniería Sanitaria y Ambiental, La Habana, pp 152–157

Fensterer C, Scholz D, Hoffmann D, Mangini A, Pajón JM (2010) ^{230}Th/U-dating of a late Holocene low uranium speleothem from Cuba. In: Publication in the YSM proceedings issue in IOP conference series: earth and environmental science (EES), 9, 012015, 6 pp (http://iopscience.iop.org/1755-1315/9/1/012015)

Ford DC (1988) Characteristics of dissolution cave systems in carbonate rocks. In: James NP, Choauette PW (eds) Paleokarst. Springer Verlag, New York, p 601pp

Garrels RM, Christ CL (1965) Solutions, minerals and equilibria. Harper and Row, New York, 437 pp

González P (1997) Estudio a escala de laboratorio de los procesos de interacción agua-roca. Tesis en opción al título de Maestro en Ciencias del Agua, CNIC, La Habana

González P, Fagundo JR, Ferrera V, Benítez G, Álvarez E (1997) Simulación química de los procesos de interacción agua-roca carbonatada en diferentes condiciones experimentales. (MOD-01). En: Memorias del III Taller de la Cátedra de Medio Ambiente, La Habana, 7pp

Herman JS, Back W, Pomar L (1986) Speleogenesis in groundwater mixing zone. The coastal carbonate aquifers of Mallorca and Menorca, Spain. In: 9th Congreso Internacional de Espeleología, Barcelona, vol 1, pp 13–15

Klimchouk BA, Ford DC, Palmer AN, Dreybrodt W (2000) Speleogenesis. Evolution of Karst Aquifers, January 2000th edn. NES, USA, p 527pp

Langmuir D (1971) The geochemistry of some carbonate ground waters in Central Pennsylvania. Geochim Cosmochim Acta 35(10):1023–1045

Miserez JJ (1973) Geochimie des eaux du karst Jurassien. Contribution physico-chimique a l'etude des alterations. These Doct. és Sciences, Neuchatel, 313pp

Molerio LF (1992) Procesos de transporte de masa en la zona o saturada de los acuíferos cársicos tropicales. In: Jaume I (ed) Libro de Comunicaciones I Taller Internacional sobre Cuencas Experimentales en el Karst, Matanzas, Cuba, Castellón (España), pp 1–15

Molerio León LF (1985) Dominios de flujo y jerarquización del espacio en acuíferos cársicos. Resumen. Simp. XLV Aniv. Soc. Espeleol. Cuba, La Habana, pp 52–53

Molerio LF, Valdés JJ (1975) Problemas y perspectivas de la investigación geoespeleológica en Cuba. Revista Tecnológica, 3/75, XIII: 34–37

Molerio LF, Guerra M, Rocamora E, Pajón JM (1999) Caracterización de los Niveles de Cavernamiento del Karst de Cuba Occidental y su Relación con las Oscilaciones del Nivel del Mar durante el Cuaternario. In: Pajón JM (ed) Paleoclima del Cuaternario Cubano: Una Caracterización Cuantitativa. Unpublished monography, Ciudad de La Habana, pp 113–170

Pajón JM (1986) Geoquímica de las Aguas Kársticas de la Sierra de los Órganos y el Pan de Guajaibón, Pinar del Río, Cuba. Tesis de Grado de Licenciatura en Química, Universidad de La Habana y Centro Nacional de Investigaciones Científicas, 226pp

Pajón JM (2007) Cambios Climáticos Abruptos en la Transición Pleistoceno-Holoceno a partir de Paleoregistros Isotópicos. Casos de estudio con espeleotemas. Segunda Convención Cubana de Ciencias de la Tierra, Geociencias´2007. Memorias en CD-Rom, La Habana, 20–23 de Marzo de 2007. ISBN 978-959-7117-16-2

Pajón JM (2009) Climates of the Past: Natural tendencies or anthropic action? IPS-Inter Press Service. Corresponsalia Cuba. Enfoques No. 21, First fortnight publication, 2008, 13pp

Pajón JM, Colectivo de Autores (1999) Monografía "Proyecto Paleoclima del Cuaternario Cubano: Una Caracterización Cuantitativa". Banco de Proyectos del Programa Nacional de Cambios Globales y la Evolución del Medio Ambiente Cubano, Ciudad de La Habana, 362pp

Pajón JM, Valdés JJ (1991) Simulación química de la disolución de rocas carbonatadas del macizo Pan de Guajaibón, Sierra del Rosario, Cuba. Revista LAPIAZ Monografía III-Valencia, Diciembre de 1991 pp 25–37

Pajón JM, Fagundo JR, Valdés JJ (1985) Relación entre el ΔpH y el RSC en las aguas kársticas de la cordillera de Guaniguanico. Libro Resumen 9 [no] Seminario Científico del CENIC, 20pp

Pajón JM, García A, Cruz C, de la Cruz A (1990) Evolución Química de las Aguas Kársticas del Arroyo Mil Cumbres (Área Protegida Mil Cumbres), S. del Rosario, Cuba (I). Libro Programa y Resúmenes, Cong. Int. 50 Aniv. de la SEC, pp 114 (25pp)

Pajón JM, Hernández I, Macle J, Ortega F (2001a) Periods of wet climate in Cuba: evaluation of expression in karst of Sierra de San Carlos. In: Interhemispheric climate linkages (Present and past interhemispheric climate linkages in the Americas and their societal effects). Chapter: Long-term climate variability. Academic Press, pp 217–226

Pajón JM, Colectivo de Autores (2001b) Monografía "Proyecto Reconstrucción paleoclimática y paleoambiental del Pleistoceno Tardío-Holoceno para Cuba Occidental". Banco de Proyectos del Programa Nacional de Cambios Globales y la Evolución del Medio Ambiente Cubano, Ciudad de La Habana, 273pp

Pajón JM, Curtis J, Tuphope S, Metcalfe S, Brenner M, Guilderson T, Chilcot C, Grimm E, Hernández I (2006) Isotope records from a stalagmite from Dos Anas Cave in Pinar del Río Province, Cuba. Paleoclimatic implications. In: CD Rom fifth international symposium on nuclear and related techniques-NURT- 2006. ISBN 959-7136-42-2

Pajón JM, Mangini A, Fensterer C, Schroeder-Ritzrau A, Scholz D, Otero V, García G, Chávez O (2009) Monitoreo químico-físico e isotópico de las aguas de flujo hipodérmico de la cueva Dos Anas (Sierra de San Carlos) y la cueva Antorcha (Sierra de Quemados), Sierra de los Órganos, Pinar del Río, Cuba. Informe Científico-Técnico. ICAN. Universidad de Heidelberg, Alemania, 45pp

Pajón JM, Mangini A, Fensterer C, Schroeder-Ritzrau A, Scholz D (2009) Dripwaters monitoring in cave systems of western karstic area of Cuba. A preliminary contribution to understanding isotopic composition of speleothems, karst hydrochemistry and kinetic of carbonate dissolution. Part I. Interdynamik project Caribclim – Heidelberg – Havana Project. Informe Científico-Técnico (ICT). Banco de Proyectos del ICAN, 20pp

Picknett RG (1964) A study of calcite solution at 10º C. Trans Cave Res Gr G B 7(1):40–50

Picknett RG (1972) The pH of calcite solution with and without magnesium carbonate present, and the implications concerning rejuvenated aggressiveness. Trans Cave Res Gr G B 14(2): 141–150

Picknett RG (1973) Saturate calcite solutions from 10 to 40º C. A theoretical study evaluating the solubility product and other constants. Trans Cave Res Gr G B 15(2):67–80
Plummer NL (2002) Geochemical and isotopic studies of paleowaters and recent recharge in parts of the Floridian aquifer system of Florida and Georgia. In: Martin JB et al. (eds) Hydrogeology and biology of post-paleozoic carbonate aquifers. Proceedings of the symposium karst frontier: Florida and related environments, Gainesville, Florida, pp 69–72, 6–10 Mar 2002
Plummer LN, Wigley TML (1976) The dissolution of calcite in CO_2 saturate solution at 25°C and atmosphere total pressure. Geoch et Cosmoch Acta 40:191–202
Plummer LN, Wigley TML, Parkhurst DL (1978a) The kinetics of calcite solution in CO_2 – water systems at 5°C to 60°C and 0.0 to 1.0 atm CO_2. Am J Sci 278:179–216
Plummer LN, Wigley TML, Parkhurst DL (1978b) The kinetics of calcite solution in CO_2 – water systems at 5°C to 60°C and 0.0 to 1.0 atm CO_2. Am J Sci 278:179–216
Rauch H, White WB (1977) Dissolution kinetics and carbonate rocks. 1 – effects of lithology on dissolution rate. Water Res 3(2):381–394, EE.UU
Richard DT, Sjöberg EL (1983) Mixed kinetic control of calcite dissolution. Am Jr Sci 283: 815–830
Roques H (1964) Contribution a l' etude statique et cinetique des systemes gas carbonique-eau-carbonate. Thése Ann de Spéléol 19(2):255–484
Roques H (1972) Sur une nouvelle methode graphique d' etude des eaux naturelles. Ann Speleol 27(1):79–92
Roques H (1973a) Etude expérimentales de la disolución des calcaries por une eau chargés de CO_2. Ann spéléol 23(1):155–167
Roques H (1973b) Sur la repartition du CO_2 dans les Karsts (2e Memorie). Ann de Speleol 18(2): 141–184
Sjoberg EL, Rickard DT (1984) Temperature dependence of calcite dissolution kinetics between 1 and 62 °C at pH 2,7 to 8,4 in aqueous solution. Geochim Cosmochim Acta 48:485–493
Stchouzkoy-Muxart T (1972) Contribution a l' etude de courbes de solubilite de la calcite dans l' eau en presence d' anhidride carbonique. Ann de Speleol 27(3):465–478
Stumm W, Morgan JJ (1981) Aquatic chemistry. An introduction emphasizing chemical equilibria in natural waters. Wiley Interscience, New York, p 583pp
Weyl PK (1958) The solution kinetics of calcite. J Geol 66:163–176
White WB (1977) The role of solution kinetics in the development of karst aquifers. In: Tolson JS, Doyle FL (eds) Karst hydrogeology. International Association Hydrogeology, Memoir, vol 12, pp 503–517
White WB (2000) Dissolution of limestone form field observation. In Klimchouk BA, Ford DC, Palmer AN, Dreybrodt W (eds) Speleogenesis evolution of karst aquifers (January 2000 edition). National Speleological Society, Huntsville, pp 149–155
Wigley TML (1973) Chemical evolution of the systems calcite-gypsum-water. Can J Earth Sci 10:306

Chapter 27
"MODELAGUA": An Interactive Program of Inverse Mass-Balance Model for Geochemical Study: An Example of its Application in Aguascalientes, Mexico

P. González, J. Fagundo-Sierra, J.R. Fagundo, M. Suárez, C. Melián, A. Cortés, J.A. Ramos, and R. Hernández

Abstract With the reference of interactive programs such as BALANCE NETPATH and PHREEQCI, for modelling net geochemical mass-balance reactions between an initial and final water along a hydrologic flow path, which also computes the mixing proportion of two initial waters and net geochemical reactions that can account for the observed composition of a final water, an interactive program of inverse model has been developed (MODELAGUA) that not only allows starting from well-known data of the chemical composition of the water and the rock to identify in a quantitative way the geochemical reactions that give origin to this composition, but also allow an analysis of mixture of waters and net geochemical reactions that can account for the observed composition of a final water, making use of a natural tracer whose geochemical behavior allows them to be used as conservative ions. In this work MODELAGUA is presented, as an interactive program of inverse model mass-balance and an example of its application.

P. González (✉) • J.R. Fagundo
Chemistry Faculty, Havana University, Zapata between G and Carlitos Aguirre, Vedado, PC 10400, Havana
e-mail: patricia_gonzalez@fq.uh.cu

J. Fagundo-Sierra
ALIVE Company, F. D. México

M. Suárez • C. Melián
Institute for Applied Sciences and Technology, Havana, Cuba

A. Cortés
Geophysics Institute, National Autonomous University of Mexico, F. D. México

J.A. Ramos
Potosine Institute of Scientific and Technological Research (IPICYT), San Luis Potosí, México

R. Hernández
Geology Faculty. University of Pinar del Río "Hermanos Saíz", Pinar del Río, Cuba

H. Farfán González et al. (eds.), *Management of Water Resources in Protected Areas*, Environmental Earth Sciences, DOI 10.1007/978-3-642-16330-2_27,

Keywords Geochemical mass-balance reactions • Modelling • Groundwater • Aguascalientes • Mexico

27.1 Introduction

The geochemical models consist of the application of physical-chemical principles to the interpretation of hydrogeochemical systems. This methodology has been developed according to two approaches: (a) the inverse one (mass-balance) that uses a well-known data of the composition of the waters and the rocks with the objective of identifying in a quantitative way the geochemical reactions that give origin to this composition; and (b) the direct one that in the basis of some well-known initial conditions on the rock-water system, it predicts the characteristics of the resulting solution of the performance of some hypothetical reactions. Both approaches have some uses and intrinsic limitations that force their use under the most appropriate conditions (Gimeno and Peña 1994; Martínez et al. 2000).

For systems with appropriate chemical, isotopic, and mineralogical data, the inverse approach of speciation and mass-balance provides the most direct means to determine models of geochemical reaction quantitatively. For systems with inadequate or nonexistent data, the model of reaction rules provides a method, in the beginning, of prediction of geochemical reactions. In some cases it is useful to combine both simulation types.

BALANCE (Parkhurst et al. 1982) is a program to calculate mass transfer for reactions in groundwater. It defines and quantifies reactions between groundwater and minerals; calculates the mass transfer necessary to account for a change in composition between two water samples; and also models the mixing of waters, redox reactions and isotopic composition. The program NETPATH (Plummer et al. 1991, 1994) is an interactive program for modelling net geochemical mass-balance reactions between an initial and final water along a hydrologic flow path, very similar to BALANCE. The main advantage of the system is that every possible geochemical mass-balance reaction model is examined between selected evolutionary waters for a set of chemical and isotopic constraints, and a set of plausible phases in the system (Parkhurst et al. 1982; Plummer et al. (1991)). On the other hand PHREEQCI expands on previous approaches by the inclusion of a more set of mole-balance equations and the additional of inequality constrain that allow for uncertainties in the analytical data (Parkhurst 1997).

An interactive program of inverse model has been developed (MODELAGUA) that not only allows starting from well-known data of the chemical composition of the water and the rock to identify in a quantitative way the geochemical reactions that give origin to this composition, but also allows the analysis of mixture of waters and net geochemical reactions that can account for the observed composition of a final water, making use of natural tracer whose geochemical behavior allows them to be used as conservative ions. In this work is presented MODELAGUA, as an interactive program of inverse model mass-balance and an example of its application.

27.2 Description of the MODELAGUA

27.2.1 Modelling Procedure

The program MODELAGUA has the objective to determine the geochemical processes that originate the chemical composition of natural waters, by means of mass-balance models and mixture analysis, also allowing the realization of Stiff graphics and the determination of hydrogeochemical patrons for the classification and determination of the type of water.

Due to these limitations to make the mass-balance correctly, it is necessary to know the lithology type to which the aquifer belongs, to select only the reactions that are logical from the geochemical point of view and to revise the validity of the results with kinetic and thermodynamic approaches.

The pattern of mass balance allows to relate a study sample with a reference and to establish the geochemical processes that have taken place. As it was already pointed out, the samples selected as references and samples should correspond with the same flow line in the aquifer; therefore, the difference of mass between both samples will be due to the geochemical processes in the region.

As the data correspond to concentration values and not to mass values, it is necessary to keep in mind the variation that can be due to concentration phenomenon, for example, the plant evapotranspiration. For this reason the concepts of concentration factor and preservative ion are introduced.

- ***Conservative ion:*** Compound that does not participate in any geochemical processes. Generally, the ion chloride can be used as a preservative ion.
- ***Concentration factor:*** Factor that quantifies the concentration processes can be determined easily starting from the selection of a preservative ion. If this factor is similar to 1, it means that there are no concentration phenomena taking place.

In comparison with commercial programs that are based on mass-balance models such as BALANCE, NETPATH and PHREEQC, MODELAGUA has the advantage of being supported on Windows which provides greater speed and easiness of operation; the data may be entered within the MODELAGUA environment or can be imported in other formats (dbf, txt, excel and mdb), and the incorporation of a new method of recognition of hydrogeochemical patrons, as well as the optimization of the calculation algorithm in the mass-balance, and in the mixture analysis through the use of natural tracers whose geochemical behavior allows them to be used as preservative ions.

27.2.2 Calculation of Hydrochemical Patterns

The classification by means of hydrochemical patron is of great utility because it allows the grouping of samples with different chemical composition and to analyze possible relationships among them; it also provides qualitative information of the processes that can influence the composition of the waters.

For these reasons it is included in the program MODELAGUA, intending as the first step in the treatment of the data, as it constitutes an important guide for the selection of the reactions when carrying out the model by means of mass balance or mixture analysis.

For the determination of the hydrochemical patterns corresponding to a study sample, the program makes the comparison of its chemical composition with each one of the values of the different present patterns. For this it takes a distance approach, the sum of the quadratic difference among the value of concentration of each ion (Ci) and the concentration value corresponding to this ion in the pattern that is analyzing (Cpi), according to the expression (Eq. 27.1):

$$D_p = \Sigma_i\left(\left(C_i - C_{pi}\right)^2\right) \tag{27.1}$$

where Ci and Cpi are expressed in part by ten of meqv/L, i takes values from 1 until the number of majority ions and p takes values from 1 until the number of existent patterns. The calculation is carried out for each patron ***p*** and one is selected with minor distance to the (Dp). The values of part by 10 of meqv/L of the different patterns do not need to be extracted of a chart; however, they are generated in their own program by means of iterative cycles.

27.2.3 Mass Balance

If two waters are connected in one line of flow, it should be expected that both possess the same chemical composition. If it is not in this way, it is clear that during the trajectory of the first sample up to the second, geochemical processes that modified their composition, have occurred. This composition difference or Ionic Delta can be used to determine the processes that take place in the region. The program MODELAGUA also considers the possibility that during the trajectory of the waters, concentration variations occurred, for example, by means of the evapotranspiration phenomenon, which is the reason why the concentration factor, the variable ***F*** is introduced. This way the calculation of the ionic delta for each compound will be (Eq. 27.2):

$$\Delta C_i = C_{i(\text{sample water})} - F * C_{i(\text{references samples water})} \tag{27.2}$$

To determine ***F***, a preservative ion can be used (q), an ion that does not participate in any interaction water-rock process, where (Eq. 27.3):

$$F = C_{q(\text{sample water})} / C_{q(\text{references samples water})} \tag{27.3}$$

This way (Eq. 27.3) the program calculates ***F*** according to the selected preservative ion and substituting in Eq. 27.2, it calculates the ionic delta for the rest of the ions.

Once the value of the ionic delta well known, it can be done to the mass balance for the determination of the geochemical processes. The algorithm of calculation of the mass balance, traditionally, is based on the solution of the Eq. (27.4) system:

$$\Delta C_i = \Sigma_k (a_k * \ b_{ik}) \tag{27.4}$$

(ΔCi: ionic delta; a_k: Mass transferred in each k process (value to calculate); b_{ik}: estequiometric coefficient of each component ***i*** in each reaction k).

The program MODELAGUA adds a process of optimization to the equations system (Eq. 27.4); it introduces a new variable E_i: error made in the adjustment of each equation ***i***, this way (Eq. 27.5):

$$\Delta C_i = \Sigma_k (a_k * \ b_{ik}) + E_i \tag{27.5}$$

Then defining the total error of the balance (Et) as (Eq. 27.6):

$$E_t = \Sigma_i \left(E_i{}^2\right) \tag{27.6}$$

It is obtained (Eq. 27.7):

$$Et = \Sigma_i([\Delta C_i - \Sigma_k (a_k * \ b_{ik})]^2) \tag{27.7}$$

With the objective of finding the solution for which the error of the balance is minimum, the process of optimization of this function is carried out deriving partially regarding to each variable to calculate (a_k) and equaling to zero the obtained equations. This way the following system of equations is obtained (Eq. 27.8):

$$\partial E_t / \partial a_k = \Sigma_i (b_{ik}) * \ [\Delta C_i - \Sigma_k (a_k * b_{ik})] = 0 \tag{27.8}$$

This is then the system of equations that is solved by the program MODELAGUA. This process of optimization contributes to the advantages of the program that allows systems to solve that do not possess the exact solution and also the system of Eq. 27.8. It does not present the restriction that the number of selected processes ***k*** have to be similar to the number of ions ***i***, as it happens with the equation system (Eq. 27.4); and for this reason the program can always offer the most approximate solution for any number of reactions and compounds that are selected.

27.2.4 Mixture Analysis

For the case to be modeled, the study sample as the result of the mixture of two reference samples is necessary to determine the mixture percentage previously for each reference. In a similar way to the concentration factor, the mixture percentage can be calculated starting from the selection of a preservative ion, an ion that only owes its difference of mass to the effect of the mixture when not participating in any geochemical process. In the case it is studied, the composition of a sample as a result of the mixture of two waters, the calculation of the Ionic Delta is different, as it is necessary to know the mixture proportion, which can be calculated starting from a preservative ion.

For a preservative ion (q), the resulting composition (CR) of the mixture of two waters of composition (CA) and (CB) will be given by the Eq. 27.9:

$$CR_q = x(CA_q) + (1 - x)(CB_q) \tag{27.9}$$

Where (x) is the fraction of compound ***q*** in mixture from the sample A and (1 − x), the corresponding for the component ***q*** in mixture from the sample B.

Obtaining ***x*** from Eq. 27.9 and multiplying by 100, the mixture percent can be determined (Eq. 27.10):

$$x * 100 = 100 * (CR_q - CB_q)/(CA_q - CB_q) \tag{27.10}$$

Therefore, the difference between the two members of Eq. 27.9 will be indicative of the processes that take place, and the values in this case are taken as ionic delta (ΔCi). This way (Eq. 27.11):

$$\Delta C_i = CR_i - [x(CA_i) + (1 - x)(CB_i)] \tag{27.11}$$

Once calculated, the mixture percent $(x * 100)$ according to Eq. 27.10 for the selected preservative ion, the ionic delta can be calculated for the rest of the ions substituting the value of ***x*** in Eq. 27.11.

Once the value of the ionic delta (ΔCi) is well known, the rest of the calculation is the same as in the case of the mass balance, by means of the equation system (Eq. 27.8). To carry out the calculations, the program needs that the analytic data corresponding to the concentration of major cations and anions are introduced as study sample water or as two reference samples: Study sample water – sample of water taken in the point where the geochemical processes wanted to be determined; and References sample water–sample of water taken in points located in the same line of flow as that of the study sample. Two reference samples are only needed in the case that is desired to model to the study sample as the result of the mixture of two types of different waters.

27.3 Application

The area studied corresponds to the called Aguascalientes Valley or Graben. This area constitutes a rectangle with coordinated UTM 2′390,000–2′490,000 N and 760,000–800,000 E (Fig. 27.1). The Aguascalientes Valley or Graben is a topographical depression of about 80 km with 20 km of maximum width in their northern and southern ends, and about 10 km of minimum width in their central part. It is flanked to the east and the west by horsts or tectonic pillars that form NW-SE and NE-SW system of fractures. The filler of The Aguascalientes Valley or Graben constitute alluviums

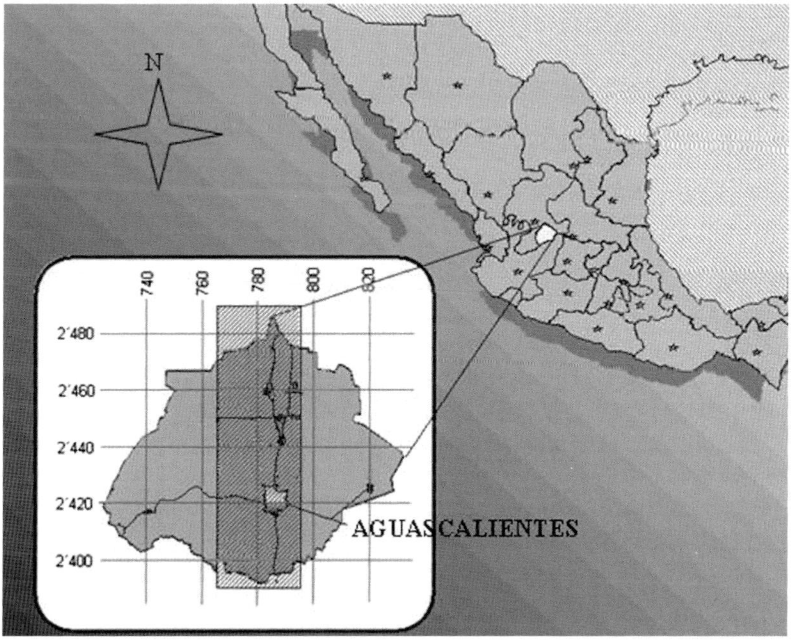

Fig. 27.1 Localization of area the study

formed by sedimentary material not consolidated. The recent alluvial material is compound, mainly, for gravels, sands and not well classified clays, which does not present stratification and in general it is not consolidated. Filler material consists of gravels and sand products of the erosion of the underlying units; in general this type of deposits stuffs the valleys and it is located in the piedmont (INAGUA-UAQ-UNAM technician report 2003).

The most important aquifer of the Aguascalientes Valley supplies a demand of almost 500 million m^3/year. The aquifer of the Valley of Aguascalientes has had an intensive exploitation in such a way that the rate of average depression 3.0 m/year, in the last decades in the urban area of Aguascalientes. Establishment problems and cracking of the land have also been related to the exploitation of the aquifer. The main areas of recharge of the aquifer of the Aguascalientes Valley or Graben come from the oriental flank and from the northern area of the Graben.

Aguascalientes aquifer is conformed in its upper part by detritus sedimentary rocks formed by alluvial materials (carried silts and deposited by the rivers and streams of the region); these alluvial materials are not consolidated, and for its grain they are of the type gravels, sands, limes and clays (Lutites) that form layers of strata of variable geometry whose thickness varies from some meters in the limits of the valley, up to 200 m in its center. According to these features, the minerals that can be found in these rocks are: Clays of kaolinite type ($Al_2Si_2O_5(OH)_4$), montmorillonite, alkaline feldspars ($NaAlSi_3O_8$, $KAlSi_3O_8$), quartz (SiO_2).

Table 27.1 Hydrogeochemical inverse mass-balance model obtained from **MODELAGUA**. Distribution of the mineral mass transformed by water-rock interaction

Name	Cl^-	HCO_3^-	SO_4^{2-}	Na^{2+}	Ca^{2+}	Mg^+	K^+
Sample AG-9	0.590	11.000	0.803	6.413	2.025	0.014	0.599
Rain water	0.160	0.200	0.000	0.210	0.040	0.038	0.010
Concentration factor	3.7						
IONIC DELTA	0.000	10.223	0.803	5.637	1.877	0.003	0.562
PROCESSES							
Albite-kaolinite	0.000	5.637	0.000	5.637	0.000	0.000	0.000
Anorthite-kaolinite	0.000	3.754	0.000	0.000	1.877	0.000	0.000
Microcline-kaolinite	0.000	0.562	0.000	0.000	0.000	0.000	0.562
Pirita-hemetite	0.000	0.000	0.803	0.000	0.000	0.000	0.000
CO_2-HCO_3^-	0.000	0.271	0.000	0.000	0.000	0.000	0.000
TOTAL	0.000	10.223	0.803	5.637	1.877	0.000	0.562
ERROR	0.000	0.000	0.000	0.000	0.000	0.003	0.000

Under the alluvial silts there are conglomerates and fractured igneous rocks. The conglomerates and the igneous rocks have a thickness from 200 to 300 m and they constitute the lower part of the aquifer. Minerals such as quartz (SiO_2), alkaline feldspars ($NaAlSi_3O_8$, $KAlSi_3O_8$), calcic plagioclases ($Ca_2Al_2Si_2O_8$), muscovite ($KAl_2[AlSi_3O_{10}][OH]_2$), micas of lithium of the lepidolite group ($KLi_{1.5}Al_{1.5}[Si_3AlO_{10}][FOH]_2$) are present. All these minerals are very insoluble, that is why the waters that drain the same one is, in general, of very low mineralization. However, some samples of the study area present high TSS, like it is the case of sample AG-9 (Beautiful Hill; Well 70), with TSS of the order of 1,085 mg/L, which is given by the high temperature (26 °C) and pH 6.50. Considering you like thermal waters because their temperature is more than 4° above the half temperature of the area (17 °C). The sample AG-9 probably originates from the deepest aquifer layer. Taking all this in consideration, the geochemical model through MODELAGUA is carried out that allows starting from the concentrations of the majority ions (Ca^{2+}, Mg^{2+}, Na^+, K^+, HCO_3^-, SO_4^{2-}, CL^-; in these samples the concentration of SiO_2 were not determined) of the sample water (in this case the AG-9) and of the representative rain water of the area to obtain the following results (Table 27.1).

The principal geochemical processes that originate the chemical composition for water sample AG-9.

$$5.64\ NaAlSi_3O_8 + 5.64\ CO_2 + 31.02\ H_2O \leftrightarrow 2.82\ Al_2Si_2O_5(OH)_4 + 5.64\ Na^+ + 5.64\ HCO_3^- + 11.28\ H_4SiO_4$$

Albite **Kaolinite**

$$1.88\ CaAl_2Si_2O_8 + 3.75\ CO_2 + 5.63\ H_2O \leftrightarrow 1.88\ Al_2Si_2O_5(OH)_4 + 1.88\ Ca^{2+} + 3.75\ HCO_3^-$$

Anorthite **Kaolinite**

$$0.56\,KAlSi_3O_8 + 0.56\ CO_2 + 3.08\ H_2O \leftrightarrow 0.28\,Al_2Si_2O_5(OH)_4 + 0.56\ K^+ + 0.56\ HCO_3^- + 1.12H_4SiO_4$$

Microcline **Kaolinite**

$$0.40\,FeS_2 + 1.5\,O_2 + 3.0\,H_2O \leftrightarrow 0.40\,Fe(OH)_3 + 0.80\,SO_4^{2-} + 1.6\,H_3O^+$$

Pirite **Hematite**

$$0.27\ CO_2 + 0.14\ H_2O + 0.07\ O_2 = 0.27\ HCO_3^-$$

27.4 Conclusions

MODELAGUA by means of an inverse mass-balance model allows the establishment of the main geochemical processes responsible for the chemical composition of the water sample, AG-9, which is the weathering of the minerals Albite and Anorthite which are abundant in the deepest aquifer layer in Aguascalientes.

References

Gimeno MJ, Peña J (1994) Basics principles of direct and inverse geochemical models. Geological Study 50(4/5):359–367

Martínez DE, Bocanegra EM, Manzano M (2000) The hydrochemical modelling as a tool en the hydrogeological study. Geologic Miner bull 111(4):110–120

Parkhurst DL (1997) Geochemical mole-balance modeling with uncertain data. Water Resour Res 33(8):1957–1970

Parkhurst DL, Plummer LN, Thorstenson DC (1982) BALANCE- A computer program for calculating mass transfer for geochemical reactions in groundwater. U.S. Geological Survey Water-Resources Investigations Res 95-4227, 143

Plummer LN, Prestemon EC, Parkhurst DL (1991) An interactive code (NETPATH) for modeling net geochemical reaction along a flow path. Geological Survey Water-Resources Investigations Report 94-4169, 130

Plummer LN, Prestemon EC, Parkhurst DL (1994) An interactive code (NETPATH) for modeling NET geochemical reactions along a flow PATH. Version 2.0. U.S. Geological Survey Water-Resources Investigations Report 94-4169, 130 p

Chapter 28
Hydrochemical Evaluation of La Gomera Aquifers (Canary Islands) and its Relationship to Garajonay National Park

M. Leal, J. Lillo, and Á. Márquez

Abstract The hydrogeological model proposed for La Gomera (Canary Islands) suggests the presence of an upper aquifer (perched groundwater bodies), a lower aquifer (General Saturated Zone) and flows through them. Many perched groundwater bodies are located under Garajonay National Park, where most of the springs are found. About 60 % of the water supply in La Gomera is covered with spring water. Therefore, if new wells are constructed and the lower and upper aquifers are truly connected, the new extractions could affect the springs of Garajonay. Hydrochemical data and multivariate statistical analyses (PCA and CA) of 120 water analyses performed by Instituto Geológico y Minero de España in 1991 show a great chemical variability of groundwater, precluding the identification of springs-groups and hydrochemical patterns, and therefore, transfer areas. Thus, it is not possible to assess the potential impact on the springs of Garajonay National Park. The factors for this hydrochemical heterogeneity could be: (1) heterogeneity of the volcanic materials; (2) influence of marine aerosols; (3) irregular rainfall; and (4) complex hydrogeological system formed by partially disconnected water bodies.

Keywords Hydrogeological model • Water supply • Volcanoes • Garajonay National Park • Canary Islands

---

M. Leal (✉) • J. Lillo
Universidad Rey Juan Carlos. ESCET, c/Tulipán s/n Dpto. Biología y Geología, 28933 Móstoles – Madrid, Spain

IMDEA Water Foundation, c/Punto Net n°4 2ªplanta, Parque Científico Tecnológico de la Universidad de Alcalá. C.P., 28805 Alcalá de Henares – Madrid, Spain
e-mail: maria.leal@urjc.es; javier.lillo@urjc.es

Á. Márquez
Universidad Rey Juan Carlos. ESCET, c/Tulipán s/n Dpto. Biología y Geología, 28933 Móstoles – Madrid, Spain
e-mail: alvaro.marquez@urjc.es

H. Farfán González et al. (eds.), *Management of Water Resources in Protected Areas*, Environmental Earth Sciences, DOI 10.1007/978-3-642-16330-2_28,

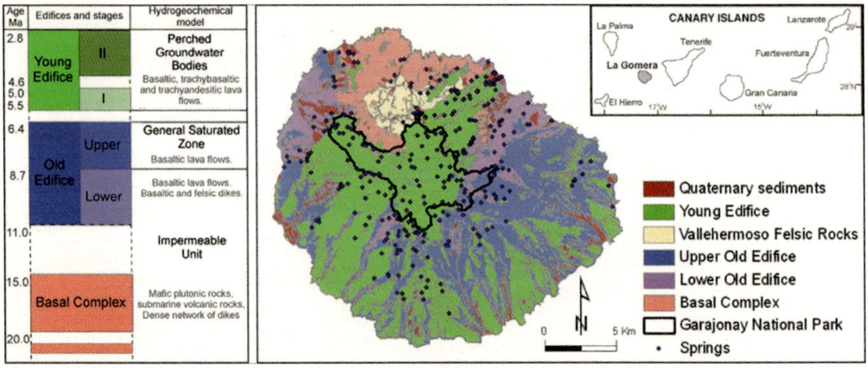

Fig. 28.1 Map of the Canary Islands with the location of La Gomera. Geologic map of La Gomera and simplified volcanostratigraphical and hydrogeological model (Modified from Ancochea et al. 2004)

28.1 Introduction

La Gomera is an inactive volcano located in the Canary Archipelago (Fig. 28.1). Although it does not present water scarcity like other Canary Islands (Custodio and Cabrera 2002), it is important to understand the groundwater flows as about 60 % of the water supply is covered with spring water, being most of the springs are in Garajonay National Park (PHIG 2003).

With that aim, several studies have been carried out for 35 years (SPA-15 1975; Porras et al. 1985a, b; APHIG 1996; PHIG 2003). Each one involved a different hydrological model. The last one was developed by APHIG (1996) and supported by the PHIG (2003). According to this model there are three hydrogeological units in the island: a unit consisting of Perched Groundwater Bodies in the horizontal basalts of the Young Edifice, a unit called General Saturated Zone in the basalts of the Upper Old Edifice and an impermeable unit consisting of the basalts of the Lower Old Edifice and the Basal Complex (Fig. 28.1). Thus, once the groundwater flowing into the horizontal basalts reaches an impermeable pyroclastic level, it cannot flow into deeper layers and starts a horizontal path. When flow cut the surface, the groundwater is discharged through a spring. However, if there is any fault in the basaltic materials, the water flows into deeper materials, reaching the General Saturated Zone. However, the transfer of water from the upper to the lower aquifer should not take place throughout the whole island due to the occurrence of a discontinuous impermeable layer between the two units.

The aim of this study is to identify those areas where there may occur a interconnection of both hydrogeological units and therefore a transfer of groundwater, in order to know if the springs of Garajonay National Park could be affected by an increase in the extraction of groundwater in the lowlands for supply.

28.2 Methodology

Previous studies made by the Instituto Geológico y Minero de España (IGME) between 1979 and 1993 in the area have been compiled to create a hydrochemical dataset. 129 springs of the 387 present in the island were surveyed. Of them, 352 water samples were analyzed (IGME 1993a, b). However, in this work only the results for 1991 are considered in order to decrease the variability affecting the water during 14 years, since our interest is mainly focused on spatial variability.

Different studies have demonstrated the usefulness of hydrochemistry (e.g. Aiuppa et al. 2003; Cruz 2003; Cruz and Amaral 2004; Prada et al. 2005; Cruz and França 2006) and statistical techniques to study volcanic and non-volcanic areas (e.g. Giammanco et al. 1998; Güler et al. 2002; Menció and Mas-Pla 2008).

Hydrochemical data analyses have been applied with the purpose of differentiating hydrochemical facies and identifying spring-waters groups. A total of 14 variables have been studied with the support of the software Aquachem 4.0: pH, electrical conductivity (μS/cm), Cl^-, SO_4^{2-}, HCO_3^-, CO_3^{2-}, NO_3^{2-}, Na^+, Mg^{2+}, Ca^{2+}, K^+, NO_2^-, NH_4^+ and SiO_2 (mg/L). In addition to the classical hydrochemical data analyses, multivariate statistical techniques as *Principal Component Analysis* (PCA) and *Cluster Analysis* (CA) were used to study hydrochemical variations and distinguish compositional groups (Richard 1998). PCA reduces the number of variables, losing as little information as possible (Davis 2002), by calculating new variables called *Principal Component* (PC) (orthogonal variables), which are linear combinations of the initial ones. The first component explains the greatest variability and successive components explains the remaining variance, each less than the previous (Jolliffe 2002). In this study the Statistical Software MiniTab 15.0 has been used for calculations using a correlation matrix. CA is a technique that groups a set of observations in a number of groups (clusters) made up of samples which are homogeneous and similar to each other and at the same time different from the samples of other groups (Everitt et al. 2001). The dissimilarity measure applied to obtain clusters has been the squared Euclidean distance. An agglomerative hierarchical clustering and the Ward's linkage method were used. The CA was performed using the software SPSS 14.0.

28.3 Results and Discussion

The Piper diagram (Fig. 28.2) shows only one dot group, which can be classified in three hydrochemical facies: Na-Cl, Na-HCO_3 and Ca-HCO_3, accounting respectively the 66.33 %, 31.63 % and 2.03 % of the total analyzed spring-waters.

Table 28.1 provides a summary of descriptive statistics. Despite the great heterogeneity of the data (with high values of standard deviation and the coefficient of variation), in general the waters have neutral values of pH (mean = 7.40), are low mineralized and the composition is dominated by Na^+ and Cl^-.

Both Cl^- and SO_4^{2-} ions are not present in the rocks of the island, so their origin must be related to rainwater and marine aerosols. However, the source of HCO_3^- ion

Chapter 29
Contamination and Protection of Surface Water Source in Czech Republic

P. Oppeltová

Abstract There have been numerous changes regarding legislative framework for water during the last years. This article deals with the problem of Kruzberk water reservoir, situated in Odra river basin (that empties into the Baltic Sea) in the northeast of the Czech Republic. The objective of this work was to evaluate the evolution tendency of reservoir water quality, determine the problematic parameters and to carry out balance evaluation of selected parameters from among the reservoir profiles in vegetation periods of the years 2006–2008. The most important profile is the profile number 1 (start of backwater) and number 5 (dam). The risk analysis based on hydrogeological exploration and area survey was carried out to protect the water quality and quantity. An optimization study and revision of protection zones were also carried out according to the valid legislation. When evaluating the concentrations, the following problematic parameters were verified: DQO – permanganate, pH, iron, manganese, calcium and magnesium, thermotolerant bacteria and Enterococcus. The regulation of activities in level 2 protection zone should positively influence water quality of the water reservoir in the future.

Keywords Reservoir water quality • Surface water protection • Contamination • Protection zones • Czech Republic

29.1 Introduction

The problem of contamination and protection of surface water sources is a theme of great importance. In the Czech Republic, most drinking water is obtained from surface and groundwater resources. Generally, groundwater resources have a lower tendency of anthropogenic pollution than surface water. However, the situation of groundwater

P. Oppeltová (✉)
Mendel University in Brno, Zemedelska 1, Brno 613 00, Czech Republic
e-mail: oppeltova@mendelu.cz

H. Farfán González et al. (eds.), *Management of Water Resources in Protected Areas*, Environmental Earth Sciences, DOI 10.1007/978-3-642-16330-2_29,

resources in soil is not perfect with regard to a drinking water supply. Therefore, the water reservoirs are used as a source of drinking water. For water quality and quantity protection it is necessary to establish protection zones. The protection zones are very important for the protection of the environment and significantly influence landscape conservation and improvement.

There have been numerous changes regarding legislative framework for water during the last years. One of the reasons for these changes was ingression of the Czech Republic to the European Union in 2004. Then, it was necessary to transpose European legislation into national law. Since both Spain and the Czech Republic are members of the European Union, they have the same legislation and therefore both countries resolve the water problems in a similar way. The most important European Directives transposed to the national legislation are: Directive 2000/60/EC and Directive 91/676/EC.

29.2 European and Czech Legislation

Directive 2000/60/EC of the European Parliament and of the Council of 23 October 2000 established a framework for Community action in the field of water policy (Water Framework Directive). By this Water Framework Directive, the European Union organizes management of surface water, continental water, transitional waters, coastal waters and groundwater to prevent and reduce its pollution, promotes sustainable water use, protects and enhances the status of aquatic ecosystems and reduces the effects of floods and droughts (The Water Framework Directive, 2009, http://www.directivamarco.es/). Its fundamental principles are:

- Hydrographic basin as the management unit, which corresponds to the unit with natural hydrological cycle;
- Cost recovery in the price of water that includes externalities;
- Achievement of good eco-biological, hydromorphological and physical-chemical status;
- Water and aquatic ecosystems recovery as a best guarantee of water quantity and quality – ecological aspect for sustainable water use;
- Reduction of groundwater pollution and elimination of dangerous substances at source.

According to this Directive, all bodies of water used for the abstraction of water intended for human consumption providing more than 10 m^3 a day as an average or serving more than 50 person (The Water Framework Directive, 2009, http://www.directivamarco.es/) must be identified.

Member States shall identify the individual river basins lying within their national territory and shall assign them to individual river basin districts. The river basins covering the territory of more than one Member State shall be assigned to an international river basin district.

The principles of the Water Framework Directive were adopted into Czech legislation by amending the Water Law 20/2004 Sb. (which came into force on January 23, 2004) and into the Drainage, sewers and public water supply law 274/2001 Sb.

Another very important European legislation is the Directive 91/676/EC – referring to protection of waters against pollution by nitrate from agricultural sources. Its principles were adopted into Czech legislation in the Regulation 103/2003 Sb. and its amendment. According to this Directive, nitrate vulnerable zones are areas where surface waters or groundwaters have nitrate concentrations of more than 50 $mg.L^{-1}$ or are thought to be at risk of nitrate contamination. The purpose of this Directive is to protect water quality by preventing high nitrate concentrations polluting ground and surface waters and especially by reducing polluting effects of the intensive cultivation and reducing the use of chemical fertilizers. It also includes regulations for wastewater treatment and good agricultural practice, such as nitrogen fertilizer use and storage, and livestock effluents. Action programs should be implemented by farmers within the nitrate vulnerable zones to prevent and reduce pollution due to nitrates from agricultural sources and should be revised every four years.

Protection zones have been defined to protect quality and quantity of drinking water sources (Water Law 254/2001 Sb.). Three water protection zones were established by previous legislation (level 1 protection zone; level 2 protection zone divided into outer protection zone and inner protection zone; and level 3 protection zone defined in surface water sources). In the actual water legislation only two levels of protection zone are defined: level 1 protection zone where more severe measures regime exists and level 2 protection zones. The actual tendency in water protection consists in determination of protection areas more defined, in result of which the water source area and the vulnerable area are not totally included in the established protection zone. It is also possible to establish more level 2 protection zones. It is of great importance, especially for agricultures, to know and observe the regulation, especially with respect to the reduction of mineral and organic fertilizers use, as well as herbicides and pesticides use. Very often it is also necessary to elaborate special programs for stock breeding prohibiting new constructions, establishment of new sewers and the use of chemicals for winter road maintaining within these zones (Oppeltová and Novák 2007).

The owners whose land is within the protection zone receive subsidy to reduce disadvantages caused due to limitation they are subjected to. In view of the large initial area of protection zones and in view of the fact that the users of water sources did not want to pay high price for the subsidy paid to affected owners, new and more reduced protection zones have been gradually delimited (Oppeltová and Novák 2007).

29.3 Kruzberk Water Reservoir and Its River Basin

Kruzberk water reservoir is situated northeast of the Czech Republic, in Odra river basin (that empties into the Baltic Sea). The water reservoir was built in Moravice River in the years 1948–1955 and is intended for accumulation of water (as a source of drinking water), Moravice River, Opava River and Odra river improvement, and floods prevention and for hydroelectrical production. This is a source of water supply for Ostrava city and surrounding municipalities, in total 200,000 inhabitants (Povodí Odry, 2009, http://www.pod.cz).

The reservoir is situated at a height of about 400 m, its river basin area is 567 km^2, the length of the reservoir is 9 km, the width is 0.5 km and its area is 280 ha. The average annual rainfall in this region is 825 mm and the average annual temperature is 7 °C, 8 °C (http://www.pod.cz). Forest area occupies approximately half of the river basin and the other half consists of agricultural land (grassland, potatoes and flax cultivation and stock breeding) (Výzkumný ústav vodohospodářský T.G. Masaryka, v.v.i., 2009, http://www.vuv.cz). A natural park is situated within the river basin where special regime must be followed. In this river basin there are also several industrial cities – Bruntál (automobiles), Bridlicna (aluminium processing) and other cities engaged in basalt extraction and plastics processing. Tourism has recently developed in this region.

There are no nitrate vulnerable zone within the basin (Výzkumný ústav vodohospodářský T.G. Masaryka, v.v.i., 2009, http://www.vuv.cz).

In the year 1967, water reservoir protection zones were established – to protect quality and quantity of drinking water. Three levels of protection zone were defined – level 1 zone, level two zones and level three zones, and the entire reservoir basin belonged to the protection zones – in total 567 km^2 (Cermak 1969; Zenaty et al. 1984).

29.4 Materials and Methods

The data on water quality and the reservoir flow values have been obtained from the state water enterprise Povodí Odry (Odra river basin). The aim of this work was to evaluate evolution tendency of water quality, determine the problematic parameters and to carry out the balance evolution of selected parameters from among the reservoir profiles (Fig. 29.1) in vegetation periods of the years 2006–2008. The analyses of all the selected water quality parameters were not performed all year long but only during the vegetation periods.

The following parameters were evaluated: iron, manganese, magnesium, calcium, carbonates, sulfates, nitrates, nitrites, ammonia, phosphates, total phosphorus, pH, oxygen, DQO – permanganate, DBO5, thermotolerant bacteria, Enterococcus bacterias and chlorophyll-alfa.

These water quality analyses were performed in five reservoir profiles (Fig. 29.1). The most important profile is the profile number 1 (start of back water) and number 5 (dam). The balance evaluation was carried out from among these profiles. The monthly weight of the water quality parameter *[1]* was calculated according to the monthly flows and concentrations and than the weight for the entire vegetation period was calculated. The quality of the water in the reservoir can be evaluated by a difference between the weight in the profile 1 and the weight in the profile 5.

Formula for calculation of monthly total substance weight (Eq. 29.1):

$$\frac{k.Q.x.c}{1000} = \text{Weight per month (kg)} \tag{29.1}$$

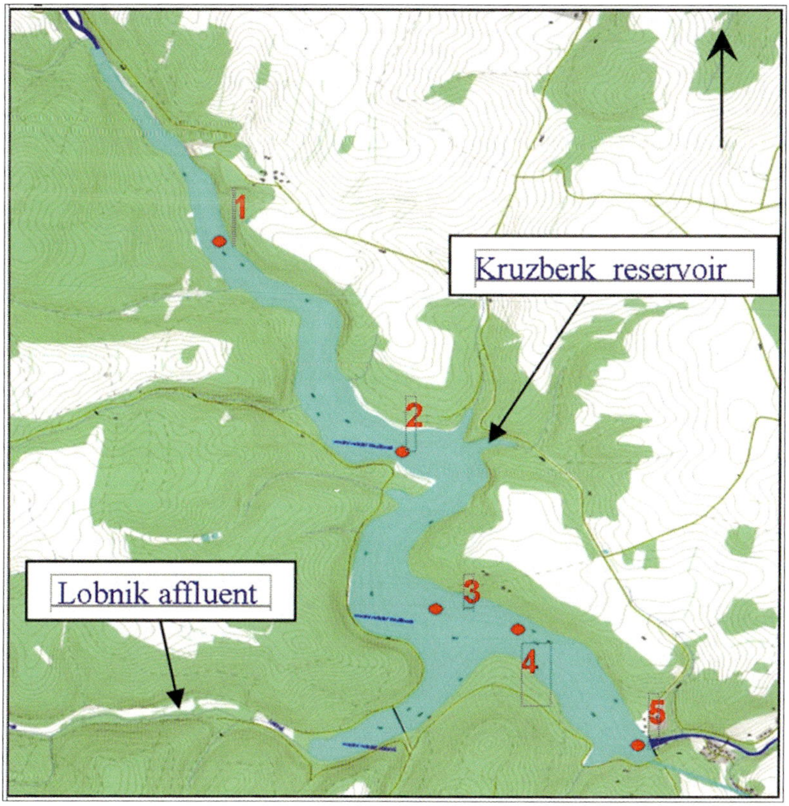

Fig. 29.1 Profiles of water quality measurement in Kruzberk reservoir

Q – Monthly flow ($m^3.s^{-1}$)
c – Substance concentration ($mg.L^{-1}$)
k – Coefficient expressing the daily amount $k = 60 \times 60 \times 24$
x – Number of days in a month

The risk analysis based on hydrogeological exploration and ground survey was carried out to protect the water quality and quantity. An optimization study and checks of protection zones were also carried out according to the valid legislation.

29.5 Results and Discussion

29.5.1 Optimization and Checks of Protection Zones

Water quality and water quantity are obviously influenced by both the natural environment and human activity. The measures to be taken in the protection zones cannot influence the aspects that depend on the natural environment. In 2007, the protection zones were revised according to the water quality results and river basin area surveys. Level 1 protection zone was defined – the basin and its surroundings (120.87 ha). Then, two individual level 2 zones were established – woodland and agricultural land surrounding the basin (247.49 ha) and the basin affluents (81.8 ha). Erosion control measures in this area must be followed and regulations with respect to the reduction of mineral and organic fertilizers use, as well as herbicides and pesticides use and special programs for stock breeding must be observed by the farmers.

29.5.2 Balance Evaluation

Important results of balance evaluation are shown below:

The limit for iron in drinking water is 0.2 $mg.L^{-1}$ (Regulation 252/2004 Sb.) and the limit for iron in A2 non-treated water is 2 $mg.L^{-1}$ (Rule 428/2001 Sb.). The legal limit for iron in drinking water was exceeded in the profile number 3 in May 2006 and in the profile number 1 in September 2008. All analyzed parameters correspond to the limits of A2 raw water.

Balance evaluation of iron:

Year 2006	5,249.09 – 3,543.45 = **1,705.64 kg**
Year 2007	6,232.59 – 6,837.67 = **–605.08 kg**
Year 2008	1,676.66 – 3,977.27 = **–2,300.61 kg**

A positive result means that the amount of iron was higher in the profile number 1 (start of back water) than in the profile number 5 (dam). The difference represents the iron retained in the reservoir. The vertical stratification of iron is produced in the reservoir. A negative result means that the amount of iron was lower in the vertical profile number 1 than in the dam. This effect can be caused by sedimentation and posterior separation of sediment from water.

The limit for manganese in drinking water is 0.05 $mg.L^{-1}$ (Regulation 252/2004 Sb) and the limit for manganese in A2 non-treated water is 1 $mg.L^{-1}$ (Rule 428/2001 Sb). The legal limit for manganese in drinking water was exceeded in the vertical profile number 2 in September 2006 and in the profile number 4 from April to August 2006. All analyzed parameters correspond to the limits of A2 non—treated water.

Balance evaluation of manganese:

Year 2006	2,736.55 – 1,852.44 = **884.11 kg**
Year 2007	2,730.50 – 2,502.46 = **228.04 k**
Year 2008	565.15 – 895.07 = **–329.92 kg**

The total amount of manganese is lower than the amount of iron. A positive result means that the sedimentation of manganese compounds occurred. A negative result shows the liberation of manganese from the water sediment.

The limit for sulphates in drinking water and in A2 non-treated water is 250 $mg.L^{-1}$ (Regulation 252/2004 Sb.). This limit was not exceeded in any case.

Balance evaluation of sulphates:

Year 2006	1,479,129 – 1,459,875 = **19,254 kg**
Year 2007	1,761,340 – 1,514,393 = **246,947 kg**
Year 2008	38,073.3 – 739,174.1 = **−1,100.8 kg**

Anthropogenic sources of sulphates are emissions containing SO_2 and SO_3 coming from the combustion of fossil fuels. A positive result means an accumulation of sulphates in the reservoir (reduction of sulphates in anaerobic environment).

The amount of sulphates in the profiles 1 and 5 are nearly equal.

The limit for nitrates in drinking water and in A2 non-treated water is 50 mg.L^{-1}(Regulation 252/2004 Sb; Regulation 428/2001 Sb.). This limit was not exceeded during the analyzed period.

Balance evaluation of nitrates:

Year 2006	294,513.00 – 322,339.60 = **−27,826.60 kg**
Year 2007	482,711.00 – 541,050.88 = **−58,339.88 kg**
Year 2008	182,292.29 – 209,221.26 = **−26,928.97 kg**

The most important source of nitrates is agricultural land. The nitrates level was always higher in the dam due to river Lobnik, an affluent that runs through agricultural land and flows into the reservoir.

The limit for phosphates in A2 non-treated water is 0.5 mg.L^{-1} (Regulation 428/2001 Sb). This limit was exceeded in the profile 3 in June 2006.

Balance evaluation of phosphates:

Year 2006	2,283.67 – 2,024.27 = 259.40 **kg**
Year 2007	2,710.27 – 3,762.53 = **−1,052.26 kg**
Year 2008	585.62 – 852.49 = **−266.87 kg**

The most important anthropogenic sources of inorganic phosphorus are the chemical fertilizers and waste waters. The organic source of phosphorus is the phosphorus from animal residues and decomposition of phytoplankton and zooplankton settling on the reservoir bottom (Pitter 2008). Phosphorus influences the reservoir water eutrophication. A positive result shows its accumulation in the reservoir (as sediment). A negative balance could be explained by phosphate supply to the reservoir, coming from fertilizers used on the agricultural land along Lobnik affluent and also from decomposition of dead organisms.

The results of the balance evaluation of total phosphorus are similar to those of phosphates – a positive result in 2006 and a negative result in 2007 and 2008.

Balance evaluation of total phosphorus:

Year 2006	5,249.09 – 3,543.45 = **1,705.64 kg**
Year 2007	6,232.59 – 6,837.67 = **−605.08 kg**
Year 2008	1,676.66 – 3,977.27 = **−2,300.61 kg**

The limit for DQO – permanganate in drinking water is 3 mg.L^{-1} (Regulation 252/2004 Sb.) and in A2 non-treated water is 10 mg.L^{-1} (Regulation 428/2001 Sb.). The legal limit DQO-permanganate in drinking water was exceeded in all the profiles and in A2 non-treated water was not exceeded in any case.

Balance evaluation of DQO – permanganate:

Year 2006	111,637.7 – 106,934.30 = **4,703.40 kg**
Year 2007	302,410.3 – 327,369.7 = **–24,959.40 kg**
Year 2008	88,582.08 – 92,330.58 = **–3,748.50 kg**

Negative results mean that there have been more organic substances in the profile 1 than in the profile 5 and vice versa.

The concentration of dissolved oxygen is an important indicator of surface water purity. Furthermore, it influences taste of drinking water (Pitter 2008). The minimum concentration of oxygen dissolved in surface waters used as drinking water is 7 mg.L^{-1}(Regulation 252/2004 Sb.). The concentrations in the reservoir evaluated vary between 4 and 14 mg.L^{-1}.

Balance evaluation of dissolved oxygen:

Year 2006	468,969.20 – 432,699.30 = **36,269.90 kg**
Year 2007	781,516.10 – 697,808.30 = **8,370.80 kg**
Year 2008	285,003.70 – 213,784.60 = **71,219.10 kg**

All results of balance evaluation are positive (the amount of oxygen in the profile 1 was always higher than in the dam). This effect can be explicated as oxygen consumption in the reservoir (decomposition of organic substances, dissasimilation, nitrification, zooplankton respiration, etc.)

The presence of thermotolerant bacteria indicates faecal contamination. Non-treated water of category A2 may contain 20,000 faecal coliform bacteria per 100 mL of water (Regulation 428/2001 Sb). This limit was not exceeded. The faecal coliform bacteria must not be detected in any 100 mL sample of drinking water (Regulation 252/2004 Sb.). This limit was exceeded in all profiles.

Balance evaluation of thermotolerant bacteria:

Year 2006	$4.60 .10^{12} - 6.91 .10^{11} = \mathbf{3.91 .10^{12}}$
Year 2007	$6.77 .10^{11} - 2.30 .10^{11} = \mathbf{4.47 .10^{11}}$
Year 2008	$1.66 .10^{11} - 1.40 .10^{10} = \mathbf{1.52 .10^{11}}$

The maximum amount of thermotolerant bacteria was always present in the summer. A positive result of balance evaluation indicates abundance of bacteria in the profile 1.

29.6 Conclusions

In the 1990s of the twentieth century, the concentrations of nitrates were very high (Zenaty 1980). Since then the concentrations have been decreasing and nowadays they do not represent a problematic indicator. On the contrary, when evaluating the concentrations, problematic parameters were verified: DQO-permanganate, pH, iron, manganese, calcium and magnesium, thermotolerant bacteria and Enterococcus. The regulation of activities in level 2 protection zone should positively influence the reservoir water quality in the future.

References

Cermak J (1969) Přehrada na Moravici u Kružberka (Kruzberkreservoir in Moravice river). Povodí Odry, podnik pro provoz a využití vodních toků, Ostrava (in Czech)
Oppeltová P, Novák J (2007) Zhodnocení problematiky ochrany vodních zdrojů v ČR v období posledních let z pohledu právního i provozního (Evaluation of problematic of drinking water sources protection in the Czech Republic in terms of law). In Zborník prednášok z X. konferencie s medzinárodnou účasťou Pitná voda 2007. Bratislava – REPRO PRINT, Bratislava, Eslovaquia, 2007, pp 63–70 (in Czech)
Pitter P (2008) Hydrochemie (hydrochemistry. VŠCHT Praga. ISBN 978-80-7080-701-9 (in Czech)
Povodí Odry (Odra river basin) [online] (2009) . http://www.pod.cz. Accessed 1 Apr 2009
The Water Framework Directive [online]. http://www.directivamarco.es/. Accessed 1 Oct 2009
Vyhláška 103/2003 Sb (National regulation 103/2003 Sb. On protection of waters against pollution by nitrate from agricultural sources) (in Czech)
Vyhláška č. 428/2001 Sb (National Regulation 428/2001 Sb.) (in Czech)
Vyhláška MZDR ČR č. 252/2004 Sb., kterou se stanoví hygienické požadavky na pitnou a teplou vodu a četnost a rozsah kontroly pitné vody (National Regulation 252/2004 Sb. On drinking water parameters and quality control) (in Czech)
Výzkumný ústav vodohospodářský T.G. Masaryka, v.v.i. [online] (2009). http://www.vuv.cz. Accessed on 7 Apr 2009
Zákon č. 254/2001 Sb. o vodách v patném znění (Water Law 254/2001 Sb.) (in Czech)
Zákon č. 274/2001 Sb. o vodovodech a kanalizacích pro veřejnou potřebu v platném znění (Drainage, sewers and public water supply law 274/2001 Sb.) (in Czech)
Zenaty P (1980) Kružberk 1955 – 1980 (Kruzberk 1955 – 1980). Vydává povodí Odry, podnik pro provoz a využití vodních toků, Ostrava, p 105
Zenaty P, Manicek J, Zubek L (1984) Povodí Odry (Odra river basin). Povodí Odry, Ostrava (in Czech)

Chapter 30
Behavior of Fecal Contamination Indicators in Waters of the Tourist Complex "Las Terrazas", Pinar del Río, Cuba

J.A. Larrea Murrell, M.M. Rojas Badía, D. Lugo Moya, and M. Heydrich Pérez

Abstract The pollution of natural waters is a serious worldwide problem due to the constant dumping of untreated or poorly treated domestic and industrial wastes. The aims of this work are to evaluate the microbiological water quality of the Tourist Complex Las Terrazas and to determine the ratio of *Escherichia coli*/fecal coliform. The samples were collected from 10 sampling stations of Las Terrazas. Fecal coliform and *E. coli* were isolated and quantified. The results show that the aquatic ecosystems of Las Terrazas do not have a high contamination degree, with values of *E. coli* within the maximum limit established by Cuban standards for recreational water and irrigation, but the fecal coliform values were slightly higher than the established standards. There is no linear correlation between the concentrations of these indicators of fecal contamination, with an average value of 0.46 for the ratio *E. coli*/ fecal coliform. These results show the good microbiological water quality of Las Terrazas and increase the information about this ecosystem located in Sierra del Rosario (Biosphere Reserve).

Keywords Microbiological water quality • *Escherichia coli*/fecal coliform • Wastewaters • Aquatic ecosystems • Las Terrazas • Cuba

30.1 Introduction

Nowadays, the pollution of natural water bodies is a serious problem, mainly in developing countries, due to the constant dumping of untreated or poorly treated domestic and industrial wastewaters, which constitutes a constant source of environmental degradation. To determine the contamination degree in these ecosystems bacterial indicators of fecal contamination are used, and among them, the most

J.A. Larrea Murrell (✉) • M.M. Rojas Badía • D. Lugo Moya • M. Heydrich Pérez
Departamento de Microbiología y Virología, Facultad de Biología, Universidad de La Habana, 25 No. 455 entre J e I Vedado, Ciudad de la Habana, Cuba
e-mail: adina@fbio.uh.cu; mayra@fbio.uh.cu

H. Farfán González et al. (eds.), *Management of Water Resources in Protected Areas*, Environmental Earth Sciences, DOI 10.1007/978-3-642-16330-2_30,

commonly used are total and fecal coliforms, although the abundance of *Escherichia coli* has been associated more with health risk compared with other coliforms (Fewtrell and Bartram 2001; Prats et al. 2008).

Nowadays, several countries, including Cuba, have a strong interest in the preservation of aquatic ecosystems for the improvement of environmental quality. For this reason, numerous studies have been performed in the Almendares River, in Havana City (Prats et al. 2006; Romeu et al. 2008) to assess the quality of the water, to contribute to the health protection of the population living in its areas.

However, this kind of study has not been carried out in the hydromineral waters from “Las Terrazas” in Pinar del Rio province. So far, in the literature appear only hydrogeographical studies made in this area by Peña (2000). Therefore, given that the hydromineral sector Las Terrazas is located in Sierra del Rosario (Biosphere Reserve) and it is an ecotourism place, it is very important to carry out microbiological studies to assess the water quality of this area located in western Cuba, which would support the feasibility of the sustainable development program that takes place in the community of Las Terrazas.

Considering these aspects, the objectives of this work are: to evaluate the microbiological water quality of the Tourist Complex Las Terrazas and to determine the *Escherichia coli*/fecal coliforms in this water.

30.2 Materials and Methods

The sampling was performed during the months of March (little rain), May and June (rainy season) in 2006 in stream ecosystems of Las Terrazas. Fecal coliform and *E. coli* were isolated and quantified.

The samples were collected from ten sampling stations, which are listed in Table 30.1. Specimens were collected in the morning and transported in sterile plastic 2 L bottles and placed in a cooler chilled. They were processed in a shorter period of time no longer than 12 h.

The concentration of coliform bacteria was enumerated by using membrane filtration technique that consists of filtering the collected samples (or dilutions of these samples) through sterile membranes of cellulose nitrate (Sartorius, with a pore size of 0.45 μm and 47 mm in diameter) using filtration equipment (Sartorius). The membranes were placed in 45 mm plates of lactose agar with Tergitol (0.095 ‰ final concentration w/v) chloride and 2,3,5-triphenyl tetrazolium (TTC) (final concentration 0.024 ‰ w/v) according to AFNOR (2001).

Colonies of yellow-orange with a yellow halo around colonies were considered a fecal coliform group of bacteria after 24 h at 44 °C. The count is expressed as colony forming units (CFU) per 100 mL of sample.

Escherichia coli were also enumerated by plate count after filtration membrane or by direct plating, depending on their abundance in the sample. For quantification a chromogenic culture medium Chromocult Agar (Merck, Darmstadt, Germany; Manafi 2000) was used. The dark blue or violet colonies obtained by the hydrolysis of substrate XGLUC (5-bromo-4cloro-3indol-ß-d-glucuronic), included in the medium,

Table 30.1 Sampling stations of Tourist Complex "Las Terrazas"

Sampling stations	Description	Latitude	Longitude
1	San Juan river before the dam	Not determined	Not determined
2	San Juan dam	Not determined	Not determined
3	Community dam	22°50′46.01″	82°56′27.09″
4	San Juan river natural pool	22°49′24.02″	82°55′35.08″
5	Nortey stream	22°50′54.09″	82°57′29.01″
6	Forest stream I	22°50′46.03″	82°59′03.07″
7	Masson stream	22°50′48.08″	82°58′26.09″
8	Forest stream II	Not determined	Not determined
9	Bayate river natural pool	22°50′22.04″	82°59′24.04″
10	Bayate river before natural pool	22°50′14.00″	82°59′26.05″

which is hydrolyzed by the enzyme ß-D-glucuronidase, after 24 h incubation at 37 °C are considered colonies of *E. coli*. The results are expressed as CFU per 100 mL of sample.

To verify the normal distribution and homogeneity of variance of the data, the Kolmogorov-Smirnov and the Cochran-Bartlett test, respectively, was carried out to the transformed data as log (x), and then the Tukey HSD test was applied to verify whether there were significant differences between the counts of *E. coli* and fecal coliform. The Pearson correlation coefficient (r) and coefficient of determination (r^2) were also calculated to assess the degree of linearity between the data in *E. coli* and fecal coliforms in Las Terrazas. For statistical calculations the statistical package Statistical 6.0 for Windows was used.

30.3 Results and Discussion

30.3.1 Microbiological Quality of the Aquatic Ecosystems of Las Terrazas

The mean of the logarithms of the fecal coliform and *E. coli* concentrations obtained from the sampling stations of Las Terrazas are shown in the Fig. 30.1. The values obtained of faecal coliform compared with the Cuban standard (Norma Cubana 1999) showed that most of the sampling stations are slightly above the maximum permissible values for recreational water (Fig. 30.1a); however, the values of *E. coli*, the most abundant fecal coliform, are within the established standards (Fig. 30.1b).

This difference is given because within the group of fecal coliform bacteria is not only *Escherichia coli*, but also in this group are species of the genera *Klebsiella*, *Enterobacter* and *Citrobacter* (Easton 1998; Marchand 2002). These genera are found in large quantities in the environment (water sources, vegetation and soil) and they are not necessarily associated with fecal contamination and do not represent an apprecia-ble risk to health (Allen 1996). However, Cuban standards express the limits of contamination in terms of fecal coliforms without considering that some members are not related to fecal pollution. This issue is not unique to our country. The United

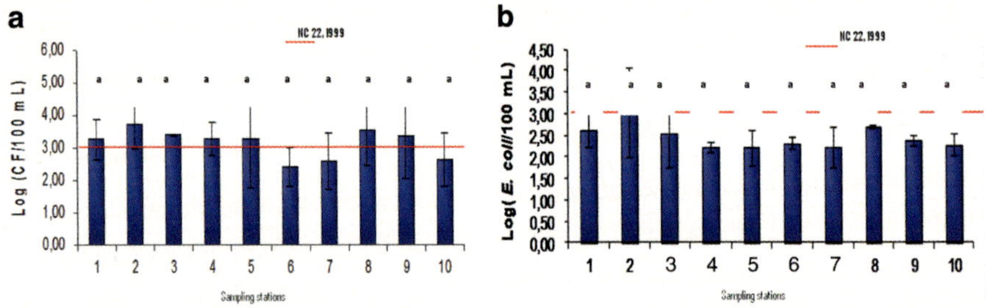

Fig. 30.1 Mean of the logarithms of the concentrations of fecal coliform (**a**) and *E. coli* (**b**) in Las Terrazas. Common letters indicate no significant differences for Tukey test ($p > 0.05$). Error bars indicate the standard deviation of three replicates

States Environmental Protection Agency (USEPA 2002) suggests that fecal coliforms are the most commonly used in the assessment of water quality. Fewtrell and Bartram (2001) argued that the abundance of *E. coli* was more associated with health risk compared with other coliforms; therefore, if this bacteria is found in large quantities in the feces of warm-blooded animals and man and does not multiply in aquatic environments, it was considered that *E. coli* would be more representative of fecal contamination in this study than fecal coliforms.

Several authors demonstrated that in tropical environmental conditions, such as high temperatures and high nutrient levels, aquatic ecosystems support the growth of *E. coli*. For example, in Hawaii (Fujioka and Shizumura 1983), Puerto Rico (Toranzos and McFeters 1997) and Sierra Leone (Wright 1982) found high concentrations of *E. coli* in the absence of known fecal sources. However, Byamukama et al. (2005) found that *E. coli* was the best indicator of fecal contamination in Uganda (African tropical country), as this bacterium was not isolated frequently in waters and soils, which suggests that it is not an autochthonous member of the studied ecosystems. This approach supports the results of this study in which *E. coli* was found in low concentrations in the waters of Las Terrazas, being representative of the microbiological quality of these waters.

To determine the existence of significant differences between the values of *E. coli* quantified at different sampling stations in Las Terrazas and between levels of coliforms, the Tukey HSD test ($p < 0.05$) was used, observing that there were not significant differences between ecosystems sampled for *E. coli* and fecal coliform, and the ten sampling stations have a similar water microbiological quality.

The waters of the aquatic ecosystems of Las Terrazas have low fecal coliform levels compared with other aquatic ecosystems at Havana City (Cuban capital) such as Quibú and Almendares River, where fecal coliform levels are between 10^4 and 10^7 CFU/100 mL (Prats et al. 2006; Romeu et al. 2008).

The presence of these coliforms in Las Terrazas waters may be due to the contribution of animal feces of wildlife, for example, birds, according to Jones and Obiri-Danso (1999) studies in the United Kingdom (UK) beaches or due to drag soil areas adjacent to the river waters as proposed Kay et al. (1999) in studies in the UK. However, in the Almendares and Quibú River, the main cause of pollution is the constant dumping of sewage from households and industry according to Prats et al. (2006) and Romeu et al. (2008) that identified the main sources of pollution in these ecosystems.

These results indicate the good microbiological water quality in Las Terrazas and have great importance because this type of study can complete the information that exists about these ecosystems located in Sierra del Rosario (Biosphere Reserve), and together with the hydrogeographic characterization by Peña (2000) constitute an integral study, which allows the analyses of the impact of man on these ecosystems and supports the benefits of the sustainable development project that is carried out in this complex tourism. Furthermore, these results provide a predictive diagnosis about the state of the aquatic ecosystems in Las Terrazas Tourist Complex to prevent them from reaching the deteriorating conditions in the rivers Quibú and Almendares.

30.3.2 E. coli/*fecal Coliforms Relationship in Las Terrazas*

Escherichia coli is a subset of fecal coliforms or thermotolerant. This bacterium is found in large numbers in the intestines of warm-blooded animals and is mainly associated with fecal contamination (Kloot et al. 2006). The United States Environmental Protection Agency (USEPA) proposed to establish as a basis for a new water quality criteria for *E. coli* (63 % of the concentration of thermotolerant coliform) to provide equivalent levels of protection for pathogens in water (USEPA 2002). The agency also suggests that the equivalent concentration of *E. coli* can be estimated using the conversion factor (ratio *E. coli*/fecal coliform) 0.63 to the concentration of fecal coliforms.

However, this value is derived from studies in which *E. coli* was enumerated with m-TEC agar, a culture-based method, which requires high temperatures (44.5 °C), which can be lethal for the stressed *E. coli* bacteria (Hamilton et al. 2005).

The specific media that are based on the detection of enzyme activity (including chromogenic or fluorogenic substrates that allow detection of the enzyme ß-D-glucuronidase) are increasingly used more compared to traditional media, because they provide a temperature optimum for growth (35–37 °C) (Leclerc et al. 2001), improving the recovery of stressed organisms. Therefore, it can be expected that the increase of recovered microorganisms increase the conversion value for *E. coli* (Hamilton et al. 2005).

That is the reason the determination of the ratio *E. coli*/FC in Las Terrazas is important, because it could allow the estimation of the concentration of *E. coli* from the fecal coliform concentration and also have a notion about the contamination level of these ecosystems.

To determine the relationship between the logarithm of the concentrations of *E. coli* and fecal coliforms, the Pearson correlation coefficient for $p < 0.05$ was calculated. Table 30.2 shows the correlation coefficient between the counts of *E. coli* and fecal coliforms obtained at Las Terrazas. The Pearson correlation coefficient was 0.4459 ($p = 0.0011$), indicating that there is no correlation between the two types of indicators.

Table 30.2 Correlation coefficient (r) between the counts of *E. coli* and fecal coliforms obtained at Las Terrazas

Organism	Fecal coliform	E. coli
Fecal coliform	1	
E. coli	0.4459* (n = 50)	1

*$p < 0.05$

This result is different from those obtained by Kloot et al. (2006) and García-Armisen et al. (2007), which reported a correlation coefficient of 0.98 and 0.94, respectively, in waters of polluted rivers, denoting a high degree of correlation between the counts of *E. coli* and fecal coliform in these kinds of ecosystems.

The fact that in this study there is no correlation between the counts of both indicators is an evidence that ecosystems of Las Terrazas did not have a high contamination level compared with the ecosystems studied by previous authors and that the origin of the two indicators is different. *E. coli* has a fecal origin and in the case of fecal coliform within which lies *E. coli*; there are other coliforms that have no fecal origin and can be found in high concentrations in aquatic environments and soil. In studies at Almendares river, Prats et al. (2006) reported that most of the concentration values of fecal coliforms and *E. coli* correlated were located on the trend line at high concentrations, obtaining a correlation coefficient of 0.75, which means that the majority of fecal coliforms were *E. coli*, which is not the case of the aquatic ecosystems of Las Terrazas in which this indicator was found in low concentrations.

To determine the existence of significant differences between the enumeration of both indicators the Tukey test was performed ($p < 0.05$). Statistical analysis result showed that there were significant differences between the counts of *E. coli* and fecal coliform counts ($p = 0.001297$), showing that fecal coliform values are higher than those of *E. coli*, which is a logical result considering that this bacterium is a subset of fecal coliforms; therefore, the relation *E. coli*/FC must be less than 1. In this study for all data collected at Las Terrazas, average ratio *E. coli*/FC was 0.46, which means that 46 % of fecal coliforms are *E. coli*; and this is further evidence of the good condition of these aquatic ecosystems.

This result is lower than that reported by USEPA (2002) and García-Armisen et al. (2007), which obtained a ratio *E. coli*/FC of 0.63 and 0.77, respectively. This difference can be given because the relation *E. coli*/FC depends on the study site and in addition to the methods applied to list these two indicators (Hamilton et al. 2005).

30.4 Conclusions

- The Aquatic ecosystems of the Tourist Complex Las Terrazas do not have a high contamination degree. The values of *E. coli* were found within the maximum limit established by Cuban standards for recreational water use and irrigation. However, the fecal coliform values were slightly higher than established standards.
- There is no linear correlation between the concentration of *E. coli* and fecal coliform in Las Terrazas, being the average value of the ratio *E. coli*/fecal coliform is 0.46, which is evidence that water ecosystems of Las Terrazas did not have a high contamination degree.

Acknowledgments This study was carried out in Tourist Complex "Las Terrazas". The authors want to thank to Fidel Hernández for his help during the fieldwork and to the CUD for the financial support of this research.

References

AFNOR – Association Française de Normalisation. Qualité de l'eau (2001) Analyses biochimiques et biologiques – Analyses microbiologiques. Tome 4. Agence Française de Normalisation, Paris

Allen M (1996) La importancia para la Salud Pública de los indicadores bacterianos que se encuentran en el agua potable. Reunión sobre la calidad del Agua Potable. CEPIS. OPS. OMS, Lima

Byamukama D, Mach RL, Kansiime F, Manafi M, Farnleitner AH (2005) Discrimination efficacy of fecal pollution detection in different aquatic habitats of a high altitude tropical country, using presumptive coliforms, *Escherichia coli*, and *Clostridium perfringes* spores. Appl Environ Microbiol 71(1):65–71

Easton J (1998) The development of a risk assessment methodology to evaluate the adverse human health effect of pathogens found in servage contaminates waters. Environmental health engineering program. University of Alabama at Birmingham

Fewtrell L, Bartram J (2001) Water quality: guidelines, standards and health. World health organization water series. IWA. Publishing, London (U.K.): ISBN: 1900222 28 0

Fujioka R, Shizumura LK (1983) *Clostridium perfringens:* a reliable indicator of stream water quality. J Water Pollut Contr Fed 57:986–992

García-Armisen T, Prats J, Servais P (2007) Comparison of culturable fecal coliforms and *Escherchia coli* enumeration in freshwaters. Can J Microbiol 53:798–801

Hamilton WP, Kim M, Thackston EL (2005) Comparison of commercially available *Escherchia coli* enumeration tests: implications for attaining water quality standards. Water Res 39: 4869–4878

Jones K, Obiri-Danso K (1999) Non-compliance of beaches with the EU directives of bathing water quality: evidence of non-point sources of pollution in Morecambe Bay. J Appl Microbiol Symp Suppl 85:101S–107S

Kay D, Wyer MD, Crowther J, Fewtrell L (1999) Faecal indicator impacts on recreational waters: budget studies and diffuse source modelling. J Appl Microbiol Symp Suppl 85:70S–82S

Kloot RW, Radakovich B, Huang X, Brantley D (2006) A comparison of bacterial indicators and methods in rural surface waters. Environ Monit Assess 121:275–287

Leclerc H, Mossel DAA, Edberg SC, Struijk CB (2001) Advances in the bacteriology of the coliform group: their suitability as markers of microbial water safety. Annu Rev Microbiol 55:201–234

Manafi M (2000) New development in chromogenic and fluorogenic culture media. Int J Food Microbiol 60:205–218

Marchand EO (2002) Microorganismos indicadores de la calidad del agua de consumo humano en Lima Metropolitana. Tesis para optar al título profesional de biólogo con mención en Microbiología y Parasitología. Universidad del Perú, Decana de América

Norma Cubana (1999) Lugares de baño en costas y en masas de aguas interiores. Requisitos higiénicos sanitarios. 1ra edn. Oficina Nacional de Normalización. Cuba

Peña B (2000) Caracterización de Sistemas Hidrominerales en el Distrito Físico Geográfico Pinar del Río. Tesis de Maestría en Geografía, Medio Ambiente y Ordenamiento Territorial. Facultad de Geografía. Universidad de La Habana. www.sld.cu/galerias/pdf/sitios/rehabilitacion-bal/tesisboris.pdf

Prats J, Larrea J, García-Armisen T, Lugo D, Rodríguez A, Servais P, Rojas N, Heydrich M et al (2006) Potential risk due to the microbial pollution in the Almendares river, Havana Cuba contribution to the education and environmental protection. Vol. 7. ISBN 959-7136-43-0

Prats J, García-Armisen T, Larrea J, Servais P (2008) Comparison of culture-based methods to enumerate *Escherichia coli* in tropical and temperate freshwaters. Lett Appl Microbiol 46(2): 243–248

Romeu B, Rojas N, Marzán R, Lugo D (2008) Necesidad de una conciencia ambientalista en torno al uso de las aguas contaminadas del río Almendares. Contribución a la Educación y la Protección Ambiental 8:209–222

Toranzos GA, McFeters GA (1997) Detection of indicator microorganisms in environmental freshwaters. In: Hurst CJ, Knudsen GR, McInerney MJ, Stetzenbach LD, Walter MV (eds) Manual of environmental microbiology. American Society for Microbiology, Washington, pp 184–194

USEPA (2002) Implementation guidance for ambient water quality criteria for bacteria (draft). EPA-823-B-003, Office of Water, Washington

Wright RC (1982) A comparison of the levels of faecal indicator bacteria in water and human faeces in a rural area of a tropical developing country (Sierra Leone). J Hyg 89:69–78

Chapter 31
Preliminary Considerations About the Bacteriological Quality of the Water Used for Human and Animal Consumption at El Tibisí, Pinar del Rio, Cuba

T. Abrantes Hernández, A.L. Ramos Hernández, F.R. Hernández Martínez, Y. Camero Labrador, S.Y. Ulloa Bonilla, and R. Novo Carbó

Abstract The use of natural waters as sources for human and animal consumption requires the evaluation of its bacteriological water quality. During the dry season, two of these water sources were sampled, following the current Cuban standard procedures for water sampling. The analytical determinations of total and faecal coliforms, *Escherichia coli*, showed values exceeding 16 colonies of E. coli/100 ml of water. This high level of contamination by pathogenic micro-organisms seems to be mainly a result of an inadequate forest management. This low quality in the water is a limiting factor for its use as a drinking water source, being a potential risk for human and wildlife health. The inappropriate hygienic habits of the dispersed population in the area composed mainly of forestry workers, along with the effects of the extensive hogs breading and other natural elements, are the basic restrictions of water quality in the study area.

Keywords Resources • Natural waters • Water quality • Contamination • Cuba

T. Abrantes Hernández (✉) • Y. Camero Labrador • S.Y. Ulloa Bonilla
National History Museum "Tranquilino Sandalio de Noda" ECOVIDA, CITMA,
Calle Martí # 202 esquina Comandante Pinares, Pinar del Río, Cuba CP: 20 100
e-mail: tamara@mhn.vega.inf.cu; ycamero@mhn.vega.inf.cu; sarita@mhn.vega.inf.cu

A.L. Ramos Hernández
Centre for Environmental Research and Services (ECOVIDA-CITMA), Km. 2½ Carretera a Luis Lazo, Pinar del Río, Cuba CP: 20 300
e-mail: aldo@ecovida.vega.inf.cu

F.R. Hernández Martínez
University of Pinar del Rio "Hermanos Saiz Montes de Oca", MES, Calle Martí final, Pinar del Río, Cuba CP: 20 100
e-mail: f.hernandez@af.upr.edu.cu; marcelo@af.upr.edu.cu

R. Novo Carbó
Botanical Garden of Pinar del Río, ECOVIDA, Pinar del Río, Cuba CP: 20 100

H. Farfán González et al. (eds.), *Management of Water Resources in Protected Areas*, Environmental Earth Sciences, DOI 10.1007/978-3-642-16330-2_31,

31.1 Introduction

From the total amount of water present in our Planet, 97 % is salt water and 3 % freshwater; out of this percentage only 0.7 % is of easy access due to the rest being frozen in the Poles and hidden in remote aquifers where its extraction is very difficult. For this reason, the actual circumstances of global water pollution affect all living beings (Gonzalez 2006).

Actually, one of every two inhabitants in developing countries suffers from a disease caused by water pollution; approximately five million persons die annually because of the use of polluted water and only 5 % of wastewater discharged in the world is treated (Pérez et al. 2006).

The term water quality is relative, the criteria as well as the standards and objectives of water quality, vary by its use. The World Health Organization (WHO) sets the limits of tolerance for substances contained in water, PHO, and the national governments set the standards at its level, varying from one country to another.

According to WHO, the quality of drinking water is a powerful environmental determinant of health. Drinking-water quality management has been a key pillar of primary prevention for over one-and-a-half centuries and it continues to be the foundation for the prevention and control of waterborne diseases. Water is essential for life, but it can and does transmit disease in countries in all continents – from the poorest to the wealthiest. The most predominant waterborne disease, diarrhoea, has an estimated annual incidence of 4.6 billion episodes and causes 2.2 million deaths every year.

There are several variants of the faecal-oral pathway of water-borne disease transmission. These include contamination of drinking-water catchments (e.g. by human or animal faeces), water within the distribution system (e.g. through leaky pipes or obsolete infrastructure) or of stored household water as a result of unhygienic handling.

As a consequence of the growing pollution and the advances of science and technology some other elements have to be considered, such as the harmful effects on human health of the use of pesticides, detergents, disinfection by-products and other organic and inorganic substances as well as protozoa, viruses and bacteria.

Water is considered a vital nutrient for wildlife. Its availability in quantity and quality is important as a limiting factor for physiological development and sanitary conditions of animals (Cseh 2003).

The transformation of the landscape and the natural habitats has disturbed the biochemical and hydrological cycles, causing erosion of thousands of tons of soil washing soil particles and sediments to the rivers, lakes and the oceans. Deforestation and chemical pollution has changed the climate in the Earth. A great number of species have disappeared and decreased its natural populations, in the last few years, due to the destruction of the natural habitats, the excess use, and the introduction of competitive species and predators (Heywood 1995).

Water has become a strategic resource, which in our opinion, has led to international confrontations and regional crises, making the water issue a consequence of the so-called environmental refugees. The main causes of the galloping shortages of

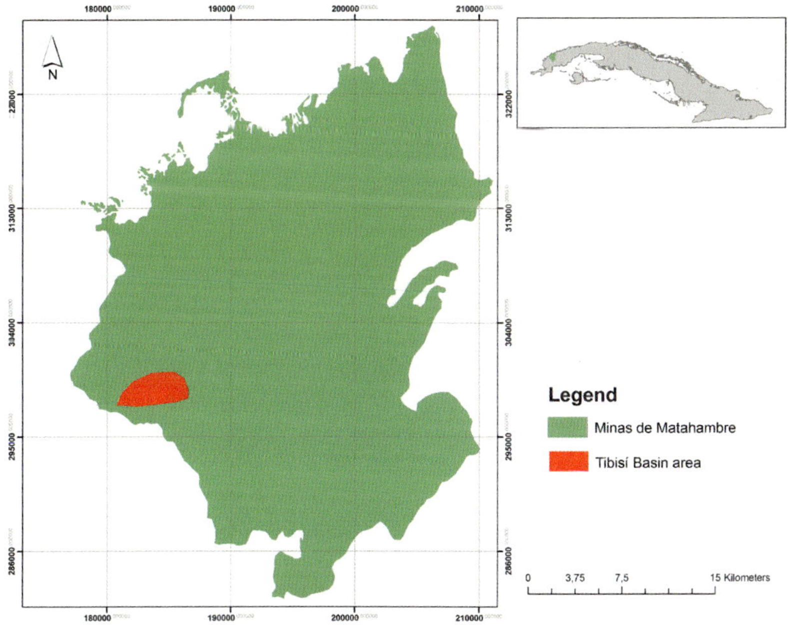

Fig. 31.1 Study area

water availability are: population growth, increasing urbanization, increasing demand for food production and the consequent increase in the agricultural sector that demands 70 % of water reserves.

Cuba has a water resource potential that amounts to 38.1 thousand million cubic meters, and actually uses almost 63 % (14 thousand million cubic meters a year) (INRH 2006). Therefore, the available potential for all uses is 1,200 cubic meters per capita of water, which shows a slight increase in the use of water resources (Batista 2002).

Water quality control is of major interest because of the presence of agents and substances rates exceeding tolerable levels that might result in damages to human health and affects natural processes and animals and plants.

The occurring waterborne diseases are associated with bad sanitary conditions and these contaminated water sources due to the discharge of industrial and domestic wastewaters. Bacteriological pollution of water originates in diarrheic outbreaks and intestinal parasitism, among other diseases. Therefore, the detailed analytical studies of water parameters are important (CITMA 2004).

31.2 Materials and Methods

During the survey, five of the natural water sources, used for human and animal consumption were sampled; these water sources are located within the area of the proposed Biosphere Reserve Sierra de los Organos in the abovementioned location

Table 31.1 Bacteriological composition of water (laboratory results)

Water samples	Dry season (March 2010) MPN coliforms in 100 ml of water	MPN Escherichia coli in 100 ml of water	Rainy season (September 2010) MPN coliforms in 100 ml of water	MPN Escherichia coli in 100 ml of water
Tibisí stream	>16	>16	>16	>16
Encinar stream			>16	>16
Loma Mala			>16	>16
Macurije river	>16	>16	>16	>16
Murria stream			5.3	5.3

of Tíbisi (Fig. 31.1). Seven (7) samples were taken. Two of them were taken during the dry season (March, 2010), one at Tíbisi stream and the other at the river Macurije. The rest of the samples were taken in September of the same year (rainy season), corresponding two of them to the same spots already mentioned, which allowed the comparison of water quality in both seasons.

The bacteriological analyses of the water samples were made at the Provincial Laboratory of Veterinary at Pinar del Río City following the standard procedures established by NORMA ISO 7251 (2009), microbiology of food and animal feeding stuffs – horizontal method for the detection and enumeration of presumptive *Escherichia coli* – most probable number technique, as well as NORMA ISO 4831 (1999), microbiology of food and animal of feeding stuffs. General guidance for enumeration of coliforms was the most probable number technique.

A comparison between the results of the bacteriological content in the samples taken at the natural sources and the reports of recent investigations available in literature, made a general study to detect potential pollution risks, as well as the proposal of new measures to be considered in the process of decision making.

31.3 Results and Discussion

The results of the analyses showed pollution levels that exceed the normal parameters for human and animal consumption (Table 31.1), both for dry season as well as for rainy season. The most probable number of *Coliform* bacteria is 100 ml (MPN/100 ml) and the MPN of *Escherichia coli* in 100 ml of water exceeded 16, except those samples taken in the natural source known as Murria, even being lower those two parameters. The results of 5.3 are a consequence of bacteriological contamination, exceeding the NORMA ISO 7251 (2009), microbiology of food and animal feeding stuffs – horizontal method for the detection and enumeration of presumptive *Escherichia coli* – most probable number technique and NORMA ISO 4831 (1999), microbiology of food and animal feeding stuffs. General guidance for enumeration of coliforms was the most probable number technique that set the limits of 2.2 MPN/100 ml.

According to Gonzalez (2006), the current circumstances of water pollution may affect the living organism.

Water is considered as a vital nutrient for livestock, if the supply is not adequate in quantity and quality, which becomes a limiting factor for sanitary and productive conditions of the livestock, as addressed by Cseh (2003).

The inappropriate hygienic habits of the dispersed population in the area, composed mainly of forestry workers, along with the effects of the extensive hog breading and other natural elements, are the basic limits of the water quality in the study area, according to the results presented in Table 31.1. It is interesting, in our opinion, the fact that even in the rainy season the bacteriological indicators of water pollution are high (higher than 16 MPN/100 ml), which is an indicator for a limited self-depuration capacity of the surface waters of Tíbisi stream, Encinar, Loma Mala and Macurije Rivers.

These results, even preliminary, are pointed to a negative impact over this territory that coincides with the muffling area of the proposed Biosphere Reserve Sierra de los Órganos, where the current economic activities have to meet special environmental requirements to guarantee the precepts of sustainable development.

31.4 Conclusions

The natural water sources used for human and animal consumption in the study area have a high level of bacteriological pollution at the locations sampled in the dry season.

The presence of a growing number of hogs due to the practice of extensive swine production as well as the impropriated hygiene habits of the dispersed population are the main causes of the water pollution in this area.

Density of swine extensive production as well as the self-depuration capacity of the water courses are two main subjects that have to be deeply studied.

References

Batista JL (2002) Utilización actual de los recursos hídricos y calidad ambiental. Noticias Mapping Iberoamérica, 5pp

CITMA (2004) Situación Ambiental de Cuba, 56pp

Cseh S (2003) Calidad de agua para consumo animal. Actas 26º Congreso de AAPA, 11pp

Gonzalez JF (2006) El Seprona lidera la sensibilización de la protección del recurso hídrico. www.guardiacivil.org/revista/result

Heywood VH (1995) Global biodiversity assessment. Cambridge University Press, Cambridge, p 74pp

INRH (2006) Gestión Integrada del Agua. Aprovechamiento de las aguas en Cuba, 6pp

March (2010) Efectos sobre la salud de la contaminación de agua y alimentos por virus emergentes humanos. Rev Esp Salud Publica [online] 79(2):253–269

NORMA ISO 4831 (2002) Microbiología de Alimentos de Consumo Humano y Animal. Guía general para la enumeración de coliformes. Técnica del número más probable, 19pp
NORMA ISO 7251 (2009) Microbiología de Alimentos de Consumo Humano y Animal – Método Horizontal para la Detección y Enumeración de Escherichia coli Presuntiva. Técnica de Número Más Probable, 21pp
Organización Mundial de la Salud (2010) Water for health. Guidelines for drinking-water quality, 6pp
Perez M, Santiesteban JA (2006) Evaluación de la calidad del agua superficial y subterránea utilizada para el consumo del ganado bovino en un municipio de provincia Habana, 15pp

Chapter 32
Presence of Illicit Drugs in Surface Waters of Protected Natural Wetlands Connected to Traditional Irrigation Systems and Urban Areas

J.A. Pascual Aguilar, V. Andreu, P. Vázquez, and Y. Picó

Abstract The Mediterranean wetlands are unique in biological diversity and they offer multiple benefits constituting a great water reserve for the planet and to produce biomass and nutrients for the trophic chain. However, the increasing human impact and the socio-economic development of the last decades have provoked important losses in these ecosystems. The work has been developed in the Natural Park of La Albufera (Valencia, Spain), which includes a coastal lagoon, marshlands, dunes and pinewoods, surrounded by rice fields in its non-urbanized part. In spite of this great ecological value, it suffers impacts derived from the high human and industrial occupation and of the hydrological contributions from the connected irrigation systems. The study has been focused on the development of a combined methodology based on environmental forensic principles to identify illicit drugs and its spatial sources and implications. Results show that rather than the pattern of population distribution the traditional irrigation system connected to location of sewage treatment plants is the way to introduce the illicit substances in the waters of the Natural Park.

Keywords Ecosystems • Wetlands • Irrigation systems • Spain

J.A. Pascual Aguilar (✉) • V. Andreu
Soil Degradation and Conservation Department, Centro de Investigaciones sobre Desertificación–CIDE, Camí de la Marjal s/n, 46470 Albal, Valencia, Spain
e-mail: juan.a.pascual@uv.es

P. Vázquez • Y. Picó
Laboratory of Food Chemistry and Toxicology, Faculty of Pharmacy, University of Valencia, Av. Vicent Andrés Estellés s/n, 46100 Burjassot, Valencia, Spain

H. Farfán González et al. (eds.), *Management of Water Resources in Protected Areas*, Environmental Earth Sciences, DOI 10.1007/978-3-642-16330-2_32,

32.1 Introduction

Mediterranean coastal wetlands are of great interest for their richness in biodiversity. They are also fragile systems because they are exposed to various human pressures such as farming systems (Readman et al. 1993) and urban sprawl (Li et al. 2010) that alter their ecological and environmental conditions.

Among other emerging contaminants, illicit drugs have also been detected in the aquatic environment. As for therapeutic substances, the main source of contamination for illicit drugs is human consumption. These substances are detectable in treated water and contaminate the receiving surface waters (Zuccato and Castiglioni 2009).

Some works have studied the presence of illicit substances in river systems and open water systems after passing through urban agglomerations and sewage treatment plants, although very little has been researched to analyze flow paths and water incorporation into permanent water bodies such as coastal lagoons. The aim of this work is the development of an integral methodology to evaluate the presence and spatial distribution of illicit drugs in surface water of the protected Natural Park of l'Albufera de Vaelncia wetland to obtain the background on how such substances travel from urban and agricultural systems to the protected area.

32.2 Study Area

The study has been applied to l'Albufera de Valencia Natural Park (Fig. 32.1), located in the east of the Iberian Peninsula. The Natural Park is surrounded by a very populated hinterland, due to the influence of the City of Valencia and its metropolitan area. Major threats come from the activities developed by a population of more than 1,200,000 inhabitants. Agriculture and intensive irrigation systems developed in and out of the limits of the Natural Park are important threats to the preservation of ecosystems and water quality.

The present Natural Park of l'Albufera is a territory of 274.4 km^2, including its marine area. Due to secular alterations, within its limits a large proportion of the land is covered by rice fields, which occupy the primitive marshland, with only a few hectares still in their natural state (Soria et al. 2002). In the continental margins of the Natural Park, intensive irrigated agricultural gardens are also found. A shallow water lagoon is located in the centre of the Natural Park. It is almost circular in form and is an area of approximately 23.7 km^2.

The hydrology of the Natural Park combines a complex human management system with natural contributions, as the water comes from the historic irrigation system, the main source of water inflows to the Natural Park. There is a very dense system of overland artificial channels for irrigation with waters mainly coming from the rivers of Jucar and Turia, which finally drains to the lake or directly to the sea.

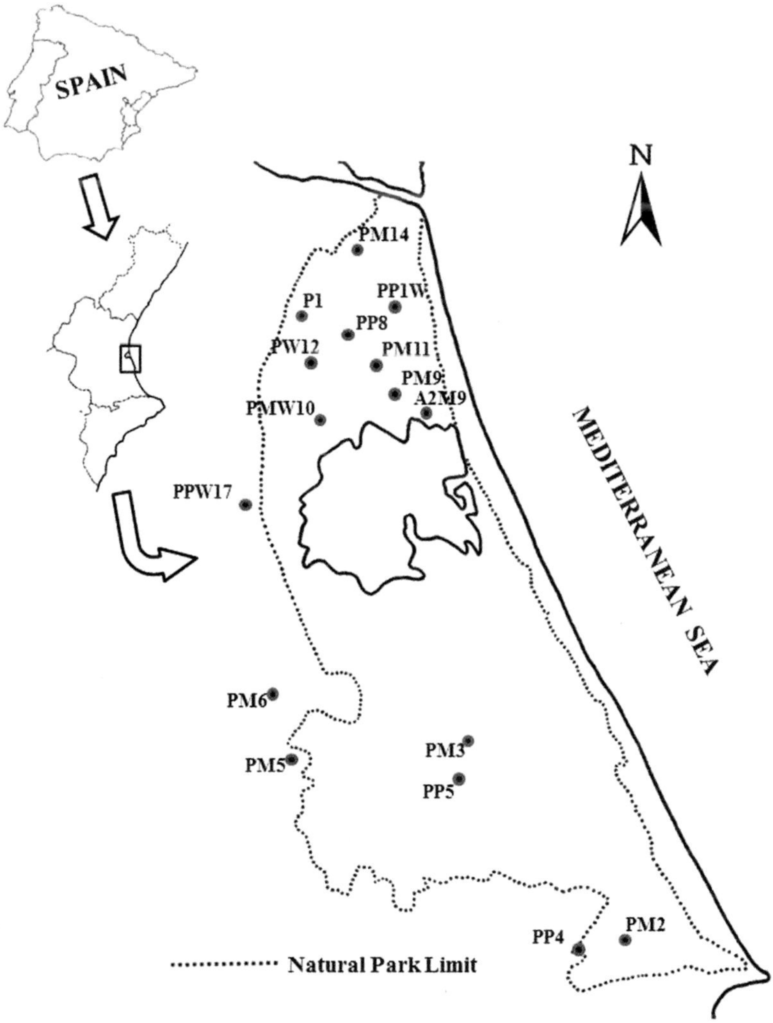

Fig. 32.1 Location of the study area and sample points setting

32.3 Methodology

Methodology is based on an environmental forensics perspective integrating different sources and data formats (Taylor 2004). It is organised around two major procedures (Fig. 32.2): analysis of water samples and a spatial analysis with Geographical Information Systems (GIS).

Initial data consisted of (1) statistic information to municipal level on number of inhabitants and population density (inhabitants by square kilometre, h/km^2) for the

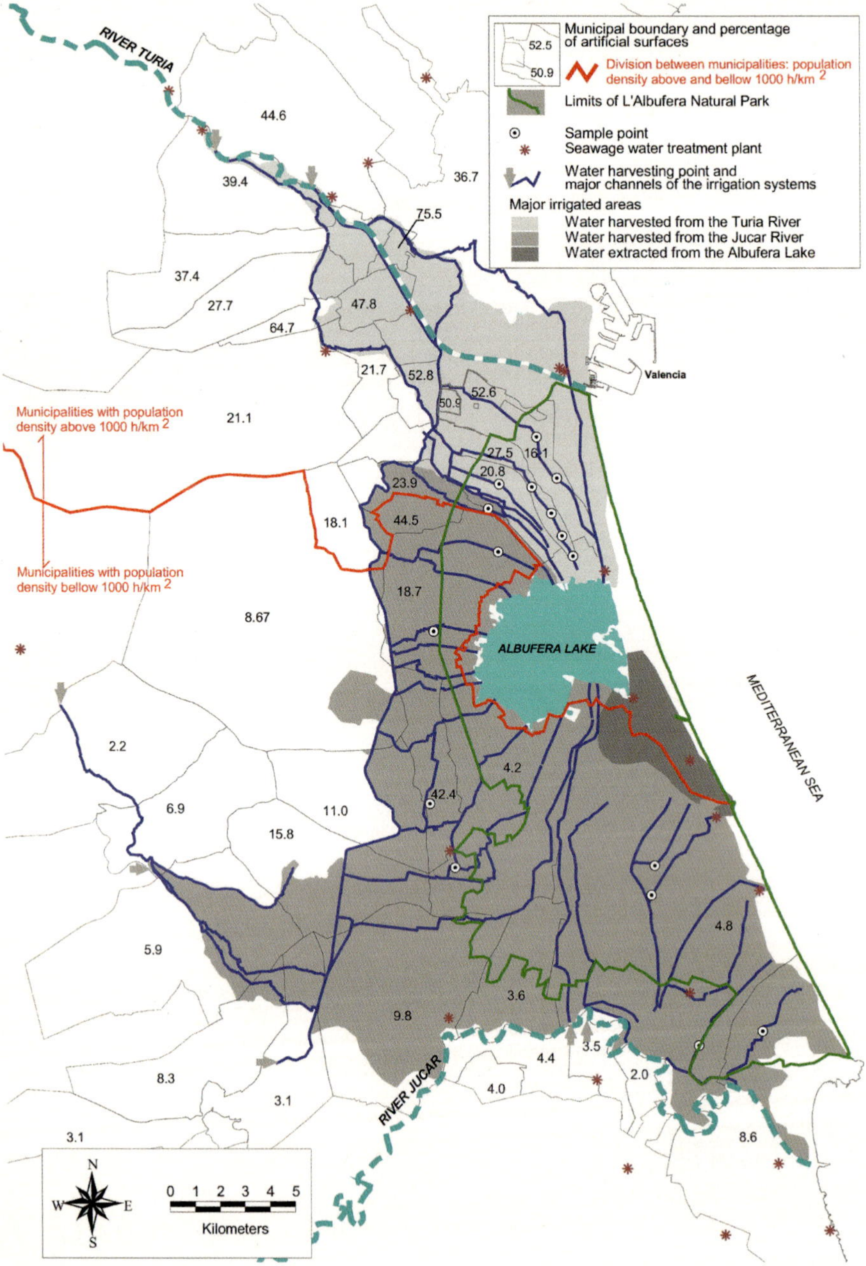

Fig. 32.2 Urban and agricultural (irrigation structure) contextualization of l'Albufera Natural Park

year 2008 provided by the Spanish institute of Statistics; (2) an updated 2008 digital layer on land cover distribution from the CORINE project (Bossard et al. 2000); (3) a digitized map with municipal boundaries; (4) a point map with the location of sewage waters treatment plants (STPs); (5) a digital layer of the traditional irrigation systems (drainage networks and areas) as stated by Hermosilla Pla (2006, 2007); and (6) 16 water samples spatially distributed over the Natural Park collected in fieldwork campaigns during the year 2008.

Water samples were further treated with a method using solid-phase extraction (SPE) and liquid chromatography tandem mass spectrometry (LC-MS/MS) for the simultaneous determination of 14 drugs of abuse and their metabolites (cocainics, amphetamine-like compounds, cannabinoids, and opiates; Vazquez-Roig et al. 2010).

From original tabular data and map layers, GIS procedures were applied to derivate new map overlays (artificial covers such as urban and industrial surfaces) and to obtain two municipal spatial indexes, the integrate percentage of urban and industrial covers and the population density.

To identify the presence of illicit drugs in traditional irrigation systems with geographical anthropogenic origins either GIS layers or results from the illicit drugs determination with the SPE and LC-MS/MS method were compared in the GIS environment.

32.4 Results and Discussion

The presence of illicit substances has been found in all out points analysed in the traditional irrigation water network of the Natural Park (Table 32.1 and Fig. 32.2), and corroborates that urban water still incorporates emerging contaminants, including illegal drugs after STPs treatment (Boleda et al. 2009; Postigo et al. 2010).

Most evidence has been found in cocainics COC and metabolites (BECG and ECGME), amphetamines (AMP and MDMA), methadone, codeine, morphine, and THC-COOH, the main metabolite of THC, which were determined in water samples at levels ranging from less than 0.14 to 78.78 ng/L. The highest concentrations were determined in the water samples collected at sample PM6, which can be related to direct spillage of residual waters from a close leisure zone.

The geographical presence of illegal substances can also be explained by a combination of water pathways and multiuse. Residual water from urban areas, after being treated in STPs, is introduced in the agricultural system for irrigation uses and finally drainage into the lake (Fig. 32.2). Traditional irrigation systems are vehicles to supply the presence of illegal substances into the Natural Park. So far, a spatial relationship between population concentration and artificial surfaces with the presence of drugs cannot be stated, although STPs drainage in the irrigation network has to be understood as the mean of illicit drugs presence in the Natural Park of l'Albufera.

Table 32.1 Presence of illicit drugs (concentration levels in ng/L) in the sample points from the l'Albufera Natural Park

	Cocainics			Amphetamine-like compounds				Cannabionids		Opiates				
	COC	BEGG	ECGME	AMP	MAMP	MDA	MDMA	THC-COOH	THC	6-ACMOR	COD	HER	MOR	MET
PM14	0.08	0.53	ND	ND	ND	ND	ND	<LOQ	ND	ND	20.8	ND	ND	ND
P1	0.34	3.01	ND	ND	ND	ND	0.18	ND	ND	ND	51.6	ND	ND	0.69
PP8	0.18	1.22	ND	ND	ND	ND	ND	ND	ND	ND	31.3	ND	ND	0.14
PP1W	ND	<LOQ	ND	ND	ND	ND	ND	ND	ND	ND	ND	ND	ND	ND
PW12	ND	<LOQ	ND	ND	ND	ND	0.76	ND	ND	ND	ND	ND	ND	ND
PM11	0.16	1.06	ND	ND	ND	ND	<LOQ	ND	ND	ND	20.1	ND	ND	<LOQ
PM9	ND	<LOQ	ND	ND	ND	ND	ND	ND	ND	ND	ND	ND	ND	ND
A2M9	0.03	1.59	ND	ND	ND	ND	0.14	ND	ND	ND	35.1	ND	ND	0.27
PMW10	ND	<LOQ	ND	ND	ND	ND	ND	ND	ND	ND	ND	ND	ND	ND
PPW17	0.08	0.89	ND	ND	ND	ND	<LOQ	ND	ND	ND	ND	ND	0.3	0.19
PM6	4.43	78.71	1.35	3.38	ND	ND	2.48	ND	ND	ND	18.69	ND	11.7	0.84
PM5	0.06	0.51	ND	ND	ND	ND	ND	ND	ND	ND	ND	ND	ND	ND
PM3	0.08	0.48	ND	ND	ND	ND	ND	ND	ND	ND	ND	ND	ND	ND
PP5	0.08	0.53	ND	ND	ND	ND	ND	ND	ND	ND	ND	ND	ND	ND
PM2	0.25	0.14	ND	ND	ND	ND	ND	ND	ND	ND	ND	ND	ND	ND
PP4	0.12	0.62	ND	ND	ND	ND	ND	ND	ND	ND	ND	ND	ND	ND

32.5 Conclusions

The combined method using water samples and spatial analysis integrated under an environmental forensic approach has been effective to detect the presence of illicit drugs in the Natural Park o l'Albufera.

There is no clear evidence of a geographical trend based on population density or concentration or urbanized areas; in that case, further work should be done to establish the effectiveness of STPs in purifying contaminated waters with illicit substances.

The analysis has proven that the water paths of inflow waters to the Natural Park are related to the overlapping of traditional irrigation systems and networks with STPs.

Acknowledgments This work has been carried out in the scope of project HUCOMED CGL2008-01693/BTE and project GCL2007-66687-C02-01/BOS, both financed by the Spanish Ministry of Science and technology.

References

Boleda MR, Galceran MC, Ventura F (2009) Monitoring of opiates, cannabinoids and their metabolites in wastewater, surface water and finished water in Catalonia, Spain. Water Res 43:1126–1136

Bossard M, Feranec J, Otahel J (2000) CORINE land cover technical guide – Addendum 2000. European Environment Agency, Copenhagen

Hermosilla Pla J (2006) Las Riberas del Xúquer: Paisaje y Patrimonio Valencianos. Generalitat Valenciana, Valencia

Hermosilla Pla J (2007) El patrimonio hidráulico del Bajo Turia: L'Horta de València. Generalitat Valenciana, Valencia

Li Y, Zhua X, Suna X, Wang F (2010) Landscape effects of environmental impact on bay-area wetlands under rapid urban expansion and development policy: a case study of Lianyungang, China. Landscape Urban Plan 94:218–227

Postigo C, López de Alga MJ, Barceló D (2010) Drugs of abuse and their metabolites in the Ebro River basin: occurrence in sewage and surface water, sewage treatment plants removal efficiency, and collective drug usage estimation. Environ Int 36:75–84

Readman JW, Albanis TA, Barcelo D, Galassi S, Tronczynski J, Gabrielides GP (1993) Herbicide contamination of mediterranean estuarine waters: results from a MED POL pilot survey. Mar Pollut Bull 26(11):613–619

Soria JM, Miracle MR, Vicente E (2002) Relations between physico–chemical and biological variables in aquatic ecosystems of the Albufera Natural Park (Valencia, Spain). Verhandlungen Internationale Vereinigung für Theoretische und Angewandte Limnologie 28:564–568

Taylor DA (2004) Environmental forensic files. Environ Health Persp 112(2):a88–a88

Vazquez-Roig P, Andreu V, Blasco C, Picó Y (2010) SPE and LC-MS/MS determination of 14 illicit drugs in surface waters from the Natural Park of L'Albufera (València, Spain). Anal Bioanal Chem 39(7):2851–2864

Zuccato E, Castiglioni S (2009) Illicit drugs in the environment. Philos Trans Roy Soc A 367:3965–3978

Chapter 33
Influence of Interactions of Surface Waters: Groundwaters on the Chemistry of Surface Waters in the River Andarax Catchment (Almería, SE Spain)

F. Sánchez-Martos, J. Gisbert Gallego, L. Molina Sánchez, and A. Pulido-Bosch

Abstract Interactions between surface water and groundwater (SW-GW) in the Andarax catchment are complex and encourage a diversity of surface waters in the three longest watercourses. The headwaters of the river Andarax flow continuously, but it is temporal in its middle reaches and ephemeral in its lower reaches; these three reaches demonstrate the effect of GW-SW dependence. Water quality is also affected by increases in nitrate and salinity in different stretches. The middle reach of the river Nacimiento carries a permanent flow due to a diffuse discharge of groundwater. In the Tabernas rambla (gulley), there is a perennial saline water flow associated with a discharge of saline groundwater, but this is not continuous over the length of the watercourse. Understanding the diversity of situations linked to GW-SW interactions is essential in this semi-arid area if its water resources are to be managed properly; therefore, these interactions need to be borne in mind when considering the water quality indicators of the surface waters.

Keywords Surface and groundwater • Andarax catchment • Water quality • SE Spain

33.1 Introduction

Groundwater-surface water (GW-SW) interactions in semi-arid areas are complex and poorly understood. Understanding their interactions helps understand the hydrological behaviour of Mediterranean rivers, as the presence of surface water in these watercourses is often related to GW-SW interactions and the typology of geological substratum. As a result, understanding and characterizing the

F. Sánchez-Martos (✉) • J. Gisbert Gallego • L. Molina Sánchez • A. Pulido-Bosch
Department of Hydrogeology, University of Almería, Campus Universitario, Almería, Spain
e-mail: fmartos@ual.es

H. Farfán González et al. (eds.), *Management of Water Resources in Protected Areas*, Environmental Earth Sciences, DOI 10.1007/978-3-642-16330-2_33,

hydrological peculiarities of each watercourse is a necessity if water resources are to be managed sustainably. Historically, these temporal water bodies have not been fully integrated into the management of the catchment. In fact, they are insufficiently dealt with in regulations such as the EU Water Framework Directive. In semi-arid regions, the impact of this can be significant and may create numerous difficulties for water resources management Sophocleous (2002). Complex aquifer geology has a profound influence on river-aquifer interactions and on the water balance of the catchment (Fleckenstein et al. 2006). Even flooding can affect the GW-SW exchanges (Doppler et al. 2007). The marked effect of the landscape on flow in the river corridors and on the mobilization of nitrates must also be considered in all these processes (Vidon and Hill 2004).

The Andarax catchment (2,265 km^2) includes parts of two singular Protected Natural Spaces: the upper slopes of the Sierra Nevada are protected as a National Park and the Tabernas Desert is a so-called Natural Enclave. In addition, the catchment includes three Sites of Community Interest that form part of the European Ecological Network "Natura 2000": the "Ramblas of Gérgal and Tabernas" and "Sur de Sierra Alhamilla" and the "Sierras of Gádor and Enix". The catchment exhibits wide climatic variability that determines its plant cover, ecosystems, water availability and the various traditional land uses, and means that the hydrodynamics of the main watercourses are highly variable from a hydrological point of view.

The present study addresses GW-SW interactions in semi-arid areas from a hydrogeological point of view. It focuses on the three longest watercourses of the Andarax catchment, each of a different type, showing a wide variety of processes where groundwater plays a significant role.

33.2 Methods

33.2.1 Study Area

A wide diversity of rocks outcrop in the Andarax catchment. In the Sierra Nevada and Sierra de los Filabres, the main outcrops are basically micaschists and quartzites, while the slopes of the Sierra de Gádor are mainly limestone-dolomites. Marls with sandy intercalations and some gypsum beds are widely distributed along the valley of the Rambla de Tabernas, while along the valleys of the Bajo (Lower) Andarax and River Nacimiento are detritic deposits. The two most important aquifers of the Andarax valley are the Sierra Gádor Carbonate Aquifer and the Detritic Aquifer of the Middle and Lower Andarax. The carbonate aquifer is composed of limestones and dolomites that outcrop along the edge of the Sierra de Gádor. The detritic aquifer extends along the central portion of the valley and consists of alluvial and delta deposits, together with deltaic sandy-silty conglomerates (Sánchez Martos 1997). The aquifer formations of the river Nacimiento are detritic. Finally, the strata along the Rambla de Tabernas are mostly

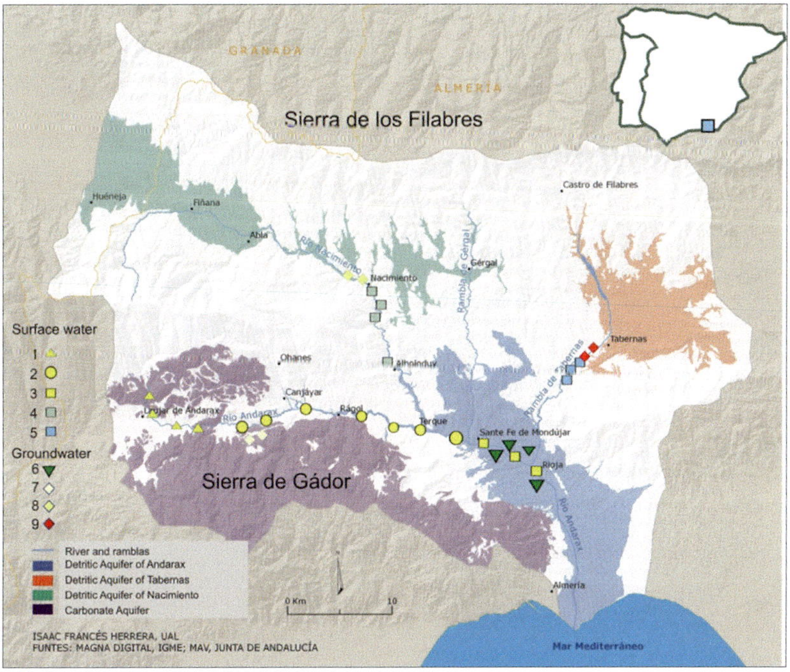

Fig. 33.1 Scheme of the Andarax catchment showing sampling points along the three watercourses studied. Surface waters: River Andarax (1: headwaters, 2: middle reach, 3: lower reach), River Nacimiento (4), Tabernas Rambla (5). Groundwaters: River Andarax Valley (6: Detritic Aquifer, 7: Carbonate Aquifer), river Nacimiento Valley (8), valley of the Tabernas Rambla (9)

impermeable, and the aquifer deposits correspond to alluvial sands and conglomerates. These alluvia cover a large surface area but their aquifer potential is restricted, given their limited depth.

33.2.2 Data

The monitoring network devised centers on the stretches of the watercourse carrying temporal surface flow, and the groundwaters in their vicinity (Fig. 33.1). Fourteen points were sampled along a 35 km stretch in the Andarax valley, with four sampling surveys made over the course of 1 year. The corresponding groundwater data considered are from the Sierra de Gádor Carbonate Aquifer and the Detritic Aquifer. Seven surface water and three groundwater points were selected in the Nacimiento valley, while in the Rambla de Tabernas 11 surface water and 6 groundwater points were sampled, with samples taken at different times. The physico-chemical parameters analyzed were: electrical conductivity, Cl^-, SO_4^{2-}, HCO_3^-, NO_3^-, Na^+, Mg^{2+} and Ca^{2+}.

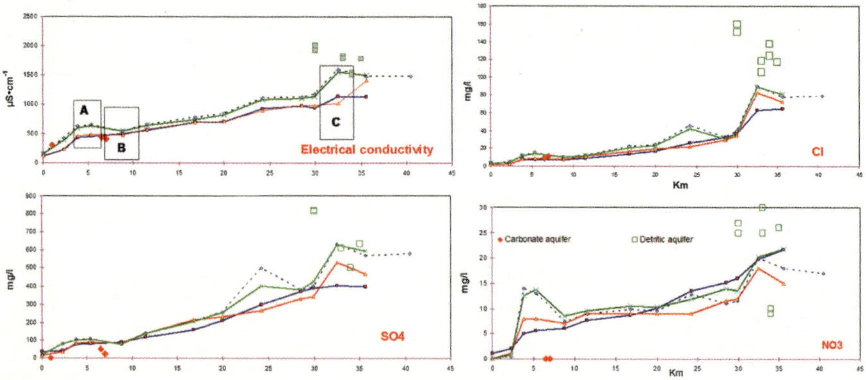

Fig. 33.2 Longitudinal evolution of electrical conductivity ($\mu S{\cdot}cm^{-1}$), Cl^-, SO_4^{2-} and NO_3^- ($mg{\cdot}L^{-1}$) in the surface and groundwaters in the Andarax valley

33.3 Results and Discussion

33.3.1 River Andarax

Groundwater in the Carbonate Aquifer has low salinity (305–441 $\mu S{\cdot}cm^{-1}$) and a calcium bicarbonate-sulphate facies. Waters in the Detritic Aquifer are more saline (1,510–2,010 $\mu S{\cdot}cm^{-1}$) and include a wide range of facies. The surface water undergoes saline enrichment along their watercourses, causing an evolution from a bicarbonate to a sulphate facies (Fig. 33.2). Rather than being gradual, this evolution includes several significant jumps in both directions.

In the headwaters (area A in Fig. 33.2), salinity is low, although there are local rapid increases in conductivity (650 $\mu S{\cdot}cm^{-1}$), sulphates (109 $mg{\cdot}L^{-1}$) and nitrates$^-$ (14 $mg{\cdot}L^{-1}$). The increase in nitrate content is marked (mean 4.5 $mg{\cdot}L^{-1}$), given the low salinity. Nitrate content of the surface waters is directly correlated with salinity ($r^2 = 0.78$) (Fig. 33.3). Nevertheless, three sampling points in the headwaters lie above the line of best fit in Fig. 33.3, indicating a local influence that pushes up NO_3^-. These three points lie in the depression of Laujar de Andarax, where traditional agricultural activities are quite intense. The peak values may relate to leaching of agricultural NO_3^-, which infiltrates the aquifer and later appears in the flow of the river Andarax (Duff and Triska 2000; Lamontagne et al. 2005).

In the area denoted B, electrical conductivity is lower than at nearby sampling points (Fig. 33.2). In this stretch, the river flows through the carbonate strata of the Sierra de Gádor. Groundwater from the carbonate aquifer flows towards the river, causing a fall in salinity, though other physico-chemical parameters are maintained throughout the year-long sampling period Sánchez-Martos et al. (2004).

Conductivity in area C rises markedly over a wide range. Sulphate concentrations here are the highest recorded in the study (Fig. 33.2). The evolution of surface flow

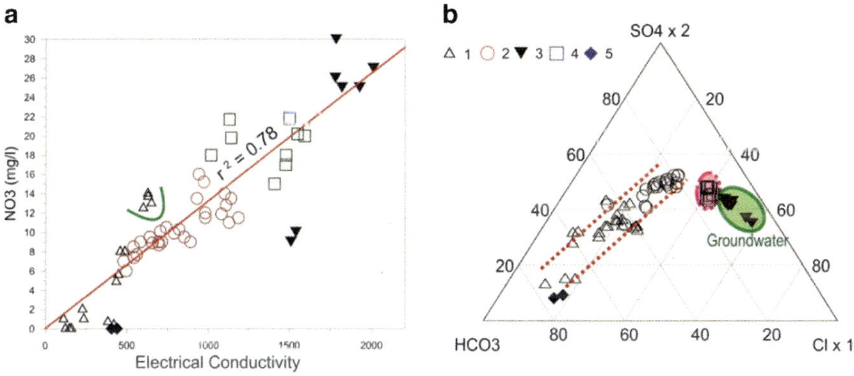

Fig. 33.3 (**a**) Nitrate content vs electrical conductivity in the river Andarax. (**b**) Triangle diagram for HCO_3^- – SO_4^{2-} × 2 – Cl^- × 12, distinguishing the surface water groups mentioned in the text (1: headwaters, 2: middle reach, 3: lower reach) and the groundwater groups (4: detritic aquifer, 5 carbonate aquifer)

in the headwaters and in the middle reach is similar (Fig. 33.3b). Sampling points are well aligned in the graph: as the percentage HCO_3^- falls, there is a corresponding increase in the percentages of Cl^- and SO_4^{2-} . In the final reach, their evolution is distinct, whereby the percentage of HCO_3^- falls and there is a gradual increase in Cl^-, while the proportion of SO_4^{2-} remains constant. The groundwater of the detritic aquifer contains higher Cl^- as a consequence of the marly deposits forming the base of the aquifer in this zone. These marly strata have been tectonically lifted, and they encourage the flow of deeper, more saline groundwater towards the surface of the aquifer. This influence is more pronounced during dry-weather periods when river flow declines and becomes discontinuous and ephemeral.

33.3.2 River Nacimiento

Groundwater was sampled from two galleries just upstream of the reach carrying perennial flow in the river. The groundwater facies is magnesium sulphate and conductivity, 1,200 $\mu S{\cdot}cm^{-1}$. The river Nacimiento flows along its entire length only after intense rainfall, though there is a small perennial flow in its middle reaches, from which samples could be taken and analysed. This water had a mixed sulphate facies and a salinity of between 1,200 and 2,100 $\mu S{\cdot}cm^{-1}$.

The perennial surface flow is fed from groundwater from the detritic aquifer in the Nacimiento valley. This flow is favoured by the presence of metamorphic rocks at the western extreme of the Sierra Nevada that form the impermeable base of the aquifer. This diffuse discharge along the riverbed causes higher ion content in the surface water than in the groundwaters sampled upstream. This diffuse inflow into the river represents a homogenization of the groundwaters, given that Cl^- and SO_4^{2-} are

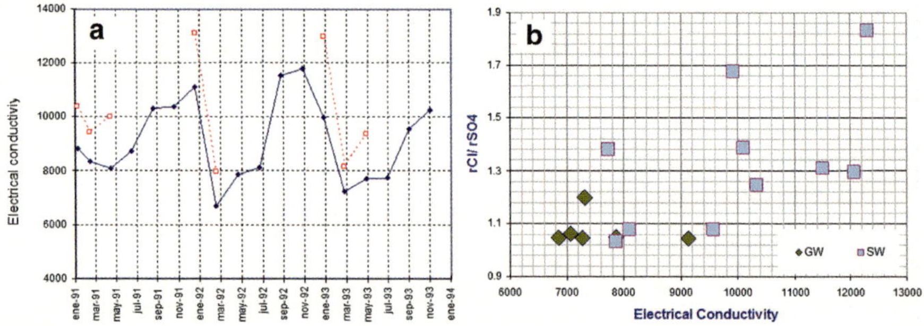

Fig. 33.4 (**a**) Evolution over time of the two surface water monitoring points in the Tabernas rambla. (**b**) Relationship rCl^-/rSO_4^{2-} vs electrical conductivity ($\mu S \cdot cm^{-1}$) for water in the Tabernas rambla

correlated ($r^2 = 0.81$) for the surface waters. The points yielding the highest nitrate concentration (10 $mg \cdot L^{-1}$) are downstream, close to Alboloduy, and probably arise as a consequence of the traditional farming activities along the river corridor.

33.3.3 Rambla de Tabernas

The groundwater comes from small, continuous surgences that drain the sandiest outcrops within the marly deposits along the margins of the watercourses. The water has a sodium chloride facies and salinity is between 6,800 and 9,100 $\mu S \cdot cm^{-1}$; thus the groundwater is more homogenous and slightly less saline than the surface waters. The surface waters have a sodium chloride facies and salinity of 7,730–13,000 $\mu S \cdot cm^{-1}$, with elevated Cl^-, SO_4^{2-} and Na^+ content. Salinity takes on a seasonal cycle, increasing during the summer-autumn and falling sharply in winter and spring (Fig. 33.4a). Despite these variations, electrical conductivity and rCl^-/SO_4^{2-} ratio are both higher than in the groundwater (Fig. 33.4b). This evolution indicates a process of cyclical saline enrichment in the watercourse over the year. Evaporation is greatest during the dry weather, therefore, increasing salinity and depositing salts that are subsequently washed downstream during the wetter seasons.

Groundwater discharge has occurred under various hydrogeological situations over time, as evidenced by the Quaternary travertine deposits that are associated with a regional fracture series (Sanz de Galdeano et al. 2008), and where there are present-day seepages of highly saline water and saline deposits.

33.4 Conclusions

The GW-SW interactions in semi-arid areas are complex and varied and give rise to diverse surface waters. The three longest watercourses of the Andarax catchment clearly exhibit the varied typology that is typical of semi-arid areas. The headwaters of the river Andarax flow continuously, while the river is temporal in its middle reach and ephemeral in the lower part. Meanwhile, the river Nacimiento is a discontinuous watercourse that has a small permanent flow in its middle reach but only flows along its whole length following intense rain. The rambla de Tabernas carries a series of small surface flows of saline water that derive from highly saline groundwater that are discontinuous in space but continuous through time.

Thus, it is demonstrated that a detailed understanding of the dependence of surface water on groundwater is essential for proper management of the associated ecosystems. The health of riverine ecosystems is heavily dependent on groundwater. This influence can work in either direction, either 'improving' or 'worsening' salinity of the surface waters, and the anthropic effects can be significant.

Lastly, the importance of understanding all the processes associated with the GW-SW interactions cannot be overemphasized in terms of the management of protected spaces in these semi-arid areas, given that the emergence of different water types favours biodiversity and needs to be considered when water quality indicators are being interpreted.

Acknowledgments This work was undertaken within the framework of project PO6-RNM-01696, financed by the Junta of Andalucía.

References

Doppler T, Franssen H, Kaiser H-P, Kuhlman U, Stauffer F (2007) Field evidence of a dynamic leakage coefficient for modelling river-aquifer interactions. J Hydrol 347:177–187

Duff JF, Triska FJ (2000) Nitrogen biogeochemistry and surface-subsurface exchange in streams. In: Jones JB, Mulholland PJ (eds) Streams and ground water. Academic, San Diego, pp 197–220

Fleckenstein JH, Niswonger RG, Fogg GE (2006) River-aquifer interactions, geologic heterogeneity and low flow management. Groundwater 44(6):837–852

Lamontagne S, Leaney FW, Herczeg AW (2005) Groundwater–surface water interactions in a large semi-arid floodplain: implications for salinity management. Hydrol Process 19:3063–3080

Sánchez Martos F (1997) Estudio hidrogeoquímico del Bajo Andarax (Almería). Tesis doctoral, Universidad de Granada

Sánchez-Martos F, Pulido-Bosch A, Vallejos Izquierdo A, Gisbert Gallego J, Fernández Cortés A (2004) Consideraciones sobre la evolución hidrogeoquímica de las aguas superficiales en el río Andarax (Almería). Geotemas 6(4):185–188

Sanz de Galdeano C, Galindo-Zaldívar J, Morales S, López-Chicano M, Azañón JM, Martín Rosales W (2008) Travertinos ligados a fallas: ejemplos del desierto de Tabernas (Almeria, Cordillera Bética). Geogaceta 45:31–34

Sophocleous MA (2002) Interactions between groundwater and surface water: the state of the science. Hydrogeol J 10:52–67

Vidon PGF, Hill AR (2004) Landscape controls on nitrate removal in stream riparian zones. Water Resour Res 40 W03201 doi: 10.1029/2003WR002473

Chapter 34
Chemical-Physical Evaluation of the Superficial Waters in Areas with Miner-Metallurgic Activity in Santa Lucia, Pinar del Río

D. Gallardo Martinez, I. Cabrera, N. Bruguera Amaran, J.A. Alonso, A. Pinto, E. Milian Milian, and A. Rodriguez Ramos

Abstract The study was carried out in the mineral field of Santa Lucia – Castellano' located to the northwest of the Pinar del Rio, Cuba, which had two main mines: Santa Lucia and Castellanos. It was in these mines where the miner–metallurgic activities like gold and silver extractions and pyrite processing were developed. The aim of the study was to diagnose the major environmental problems related to the hydrology in this area created by wastewaters that formed the acid drainage, and the contaminations with high toxicity levels. The quality of waters was corroborated, which found that the most damaged areas were: quarries, sterile deposits of the mine, and the mangrove ecosystem due to the accumulations of superficial currents and airborne transportation of fine materials. Thus, it enabled the proposal of the necessary actions to mitigate and supervise the area.

Keywords Miner-metallurgic activities • Hydrology • Ecosystems • Cuba

D. Gallardo Martinez (✉) • N. Bruguera Amaran • A. Rodriguez Ramos
Territory Delegation of the Ministry of Science, Technology and Environment, CITMA, Pinar del Rio, Cuba
e-mail: dgm@uma.vega.inf.cu; delegado@citma.vega.inf.cu

I. Cabrera • J.A. Alonso
CIPIMM, Managerial Group GEOMINSAL, MINBAS, Havana, Cuba

A. Pinto
SEGEMAR, Geologic-Miner Service, Institute of Mining Technology (INTEMIN), Buenos Aires, Argentina

E. Milian Milian
Geominera Company, MINBAS, Pinar del Río, Cuba

H. Farfán González et al. (eds.), *Management of Water Resources in Protected Areas*, Environmental Earth Sciences, DOI 10.1007/978-3-642-16330-2_34,

34.1 Introduction

Through the centuries mining has been part of the historical and economical development of numerous countries in the world. However, the social and environmental effects generated by these activities of this industry are a detriment to the different ecosystems.

The mining is shown to be an inexorable activity, depredator of the environment. However, the problem is not considered to be in the mining type process, but rather in the way it takes place, its exploitation, and the technology in the socio-political systems where it explodes.

The Matahambre region of Santa Lucia, during the year, has been the object of an intensive mining exploitation, which has caused negative consequences to the environment. The natural effect caused by the existence of numerous outcrops of sulphurous mineralization in a great part of the region lowers the action of the agents meteoric and causes the remobilization of numerous polluting elements that pass to the waters, to the soil, to the vegetation, the fauna and the atmosphere. Nowadays, these locations constitute passive miners.

The drainage of the superficial currents has been modified due to various causes. However, all the causes, such as the construction of micro-reservoirs, access roads and sewage systems; the incorporation of low law current ores, cholera associated with drainage problems; and created silt beds were associated with the previous exploitation mining activities.

In this area of extensive mining activity, the Acid Drainage of Mine or Rock, is a physical-chemical process in which the chemical oxidation of the mena occurred and in one of the biological microorganisms that catalyze the reaction of the primary sulphur participate: pyrite, sphalerite and galena in these passive miners, diminishing in a significant way the values of pH ends (<3.0), (Silverman and Lundgren 1959; Delgado Diez et al. 2009) the residual waters that pour their caudles toward the two more important rivers from the areas around to the deposits: the river "Santa Lucia" and "The Palm," respectively.

The objective of the study is to evaluate the pollution elements in the superficial currents within the mining activities of the western region of Pinar del Río, using a novel analytic method to facilitate the implementation of action for mitigating the environmental problems.

34.2 Materials and Methods

34.2.1 Selection, Preparation and Description of the Samples

The samples obtained from the study region, which were formed by the deposits of "San Lucia" and "Castellano", were selected from a wide net of sampling points according to the macroscopic characteristics of contamination that were observed, and with their corresponding coordinated GPS of localization. During the sampling process the experience of the investigators, who took samples within the mine, was kept in mind. The selected samples are described in Table 34.1.

Table 34.1 Macroscopic description of the samples (silts, salts, residual liquids, river waters and of sea) near to the deposits "Santa Lucia" and "Castellano"

Sample	Selection of samples and localization in the passive miners
M-1[a]	Residual liquid that comes from the runoff inside the own mine of "Santa Lucía"
M-2[a]	Residual liquid collected to the exit of the tunnel of the mine Santa Lucía
M-3	Silt of the glide of the reservoir that comes from the watery solution of the sample-1A of the mine Santa Lucía (X: 194,149; and: 315,153)
M-3[a]	Residual liquid that comes from the runoff on the side of the tunnel of the mine Santa Lucía
M-4[a]	Residual liquid that comes from the sterile mineral of Castellano
M-5[a]	Residual liquid of the process of lixiviación of the gold of the fewer of the hat of iron of the mine Castellano
M-7[a]	Sample of taken seawater of the channel behind the sulfometales plant (natural lagoon)
M-8[a]	Sample of seawater that comes from the jetty located behind the sulfometales plant
M-9[a]	Shows of water of the effluent of the river Santa Lucía (in situ) (X: 197,513 AND: 314,945)
M-10[a]	Shows of water of the own river Santa Lucía taken in situ (X: 197,513 AND: 314,945)

Once the samples were selected at the DCM laboratory of Mineralogy, the solids were dried until the samples were at ambient temperature. The samples were then pulverized to a grain size 100 % below a sieve measurement of 0.074 mm (200 meshes) with the objective of analyzing the corresponding chemical and mineralogical analyses. However, the analyses carried out were the determinations of the pH of the silt samples in watered solutions, and the residual liquids collected in both deposits (diluted river and seawaters of Santa Lucia).

34.2.2 Diffraction of Ray-X (RRD)

The diffractometric registrations were carried out utilizing the powder method in a team Philips PW 1,710. All the diffractometric registered according to the variant of measurement point to point, angular step of 0.050, and time of measurement in each position of 3 s. The numeric results of relative intensities and diffraction angles resulted in a continuous diffractogram with the employment of the program "Origin 7.0". The interplane distances were determined with the Ttod program developed for PC. The qualitative analysis of phases was carried out with the use of the PCPDFWIN database; version 1.30, JCPDS-ICDD/2002, compatible with Windows 98 for Office 2000.

34.2.3 Electronic Microscopy of Sweeping with Spectrometry Dispersive in Energy

Observations of the sample microstructures were carried out utilizing a Sweeping Electronic Microscope (MEB) model XL30 ESEM from the Institute INTEMIN-SEGEMAR of the Republic of Argentina. The punctual elementary chemical composition was analyzed by utilizing an energy dispersive X-ray spectroscope (EDAX). The experimental conditions are consigned in each particular spectrum.

34.2.4 Chemical and Mineralogical Analysis

The chemical analyses of the majority of the elements and their trace were determined by the acid break-up and coalition methods, utilized for the preparation of the samples. An Atomic Emission Spectrometer was used inductively with SPECTRO (Germany) Spectroflame Inductive Plasma Coupler (ICP-AES).

According to the previous geologic studies that were carried out for the Geologic Company of Pinar del Rio, today the Geominera West Company located in "Santa Lucia," the lesser of this deposit is subdivided into four bodies receiving mining that separates each other for flaws. The composition chemical average of the fundamental elements of interest (bodies 2, 3 and 4 pyrite – polymetallic with barite mineralization) is the following: lead 1.5 %, zinc 5.78 % and sulphur 28.96 % (body 2). Body 3 presents a content average of lead of 1.32 %, zinc 4.55 % and sulphur 24.49 %. Lastly, in body 4 the composition chemical average is the following: lead 2.87 %, zinc 6.09 % and sulphur 20.86 % (mentioned in Cañete et al. (2008)).

The deposit of "Castellano" on the other hand, presents a similar genesis and a geologic structure that is very similar in mineralogical composition and polimetallic, but differs partly in their mining bodies with regards to those of "Santa Lucia" for the elements: Fe, Pb, Zn and Ba (Cañete et al. 2008).

The characteristic miner graphics of the minerals, present in the samples with fewer polimetallics, were determined through refined sections, where they were analyzed with a JENAPOL polarizing microscope coupled with a Sony videocamera, model SSC-C 370 and a YS – W150 interface, utilized for analyzing images, and DIGIPAT software for the analysis of the obtained results.

34.3 Results and Discussion

34.3.1 Current Environmental Situation

The superficial waters exhibit low values of the pH (acids), in the currents that drain ores low law, areas with line deposits, and excavations or quarries of mineral extraction. The obtained values of dissolved oxygen do not reflect the existence of anoxic conditions. The observed minimum values of turbidity belong together with acid drainages of the solid residuals (choleras), as well as the quarry and the effluent of the supply spring.

In the marine area, contamination has been received from acid drainage of the laundered mineral bodies from the exploited mines "Santa Lucia" and "Castellano", which have been incorporated into the superficial currents.

34.3.2 Chemical Analysis

Cañete et al. (2008) reported on samples of silts collected along the basin of the river Santa Lucia and Palm in an investigation during the year of 2008, value anomalous of the pollutants: Fe, Zn and Pb, if these are compared with the permissible maximum limits that settle in the Cuban Norm NC 521 (2007) corresponding and that they were the following: Fe (0.9–7.3 mg/l) and Zn (18.0–481.0 mg/l) and Pb (231.0–7,408 mg/l), respectively. The authors also detected high contents of Ba in the silts of the two analyzed rivers that caused the physical-mechanical processes of haulage of the mineral barite from the primary ore to the river "Santa Lucia", a mineral that is also non-soluble in water (Kps = $1.08.10^{-10}$; 25 °C), which justifies the anomalous values of barium present then in the silts of the rivers of the area studied by the referenced authors. In a similar way, in this investigation the same behavior was observed in connection with the element barium.

The authors point out that in the waters of the basins of the river "Santa Lucia" they also reached significantly higher values of previous polluting elements: Fe (1.6–669 mg/l); Zn (3.7–407 mg/l); Pb (0.7–2.2 mg/l) that coincides with the high tenors of these elements in the samples of analyzed wastewater. Also outstanding were significantly higher results of total sulphur due to the processes of acid lixiviation of the sulphurs in the primary ores: pyrite, galena and sphalerite.

In Table 34.2 the levels of the polluting elements are shown: Zn, Pb and Fe. Also outstanding in the samples are high contents of PPI which were carried out in temperature intervals of 110–1200°C, and which justify the loss of the SO_3 associated with the sulphurs in the first place (pyrite, galena, sphalerite), sulphates (jarosite and ox – sales of Fe II, barite), the structural water of the clays and iron ox-hydroxides (both expressed as loss of structural water, H_2O^+), those that confirm the intensity of the processes SDM associated with these silts and the primary rocks of both deposits.

In Table 34.3 the results of the pH determinations are presented for the watery solutions obtained in the silts analyzed according to the Norma ASTM corresponding D-12938.

The high acidic values that were obtained from the solutions of all the silt samples are explained by the presence of ions H_3O^+ and S^{2-} dissolved inside the womb of the solid, those that when combined in watery solution and in presence of air react and give place to the sulphuric, given as acid formation in the occurrence of the processes SDM; those that according to the consulted literature (Guides for the Environmental Handling SDM in Peru 2004; Zamora 2006), they are justified for the oxidation-break-up reactions from the primary sulphurs to acid pH, through the oxidation of the pyrite with formation of iron sulphate II and ferric hydroxide, as well as the oxidation of the iron sulphate II at III, catalyzed by *Acidithiobacillus Ferrooxidant* and the reaction of the ferric sulphate with the pyrite.

The presence of micro-organisms acidophilus in the water of the Santa Lucia mine showed, at the end of 8 days a coloration change from a liquor of green to a

Table 34.2 Results of chemical analysis of polluting elements in river waters and liquid drainages of the deposits "Santa Lucia" and "Castellano"

	Concentration of the element (mg/l)								
Samples	Zn	Pb	Ba	Fe	S (total)	Mn	Ni	Mo	Sr
M-1A	6.060	64.98	<0.2	14.700	15.300	420	<0.2	<0.2	<0.2
M-2A	5.660	58.96	<0.2	13.900	11.060	220	<0.2	<0.2	<0.2
M-3A	81	62.38	<0.2	14.120	15.760	450	<0.2	<0.2	<0.2
M-4A	110	22.14	<0.2	5.030	8.550	360	<0.2	<0.2	<0.2
M-5A	<1.8	<0.4	<0.2	6.40	480	0.27	<0.2	<0.2	<0.2
M-7A	<2.1	<0.4	<0.2	1.83	970	<0.2	<0.2	<0.2	<0.2
M-8A	<1.8	<0.4	<0.2	8.80	940	<0.2	<0.2	<0.2	<0.2
M-9A	<1.7	<0.4	<0.2	4.60	85	0.30	<0.2	<0.2	<0.2
M-10A	<2.1	<0.4	<0.2	4.60	1.4	0.23	<0.2	<0.2	<0.2

Table 34.3 pH values and electric conductivity obtained in the watery solutions of the solid phases (silts) of the selected samples of "Santa Lucia" and "Castellano"

Samples	pH	Conductivity (μS/cm)
M-1	3.0	11.7
M-2	2.9	5.6
M-3	3.3	2.2
M-4	2.9	29.7
M-5	3.2	3.3
M-7	3.1	1.6
M-8	7.2	1.4
M-9	3.5	26.5
M-10	7.1	11.5

reddish brown, evidencing the presence of iron. The bacteria autotrophy, aerobic, chemiclitotrophy, mesophil, acidophil *Acidithiobacillus Ferrooxidant, and* previously well-known *Thiobacillus Ferrooxidant,* identified as AFSL1-09 with a cellular concentration in these waters of 10^5 cel/ml were isolated. The growth of the micro-organisms was continued by the analysis of the ion content Fe^{3+} present in the means of cultivation as in Karavaiko et al. (1972).

With this test and the previous corroboration, it explained the occurrence of the processes SDM in the mine of "Santa Lucia" and the important fact of the micro-organisms in this case on the phase mineral pyrite, which is the most sensitive phase to the action of the agents micro-biologics.

For the above-mentioned, the high concentrations of polluting metals detected in the silts explain the high acidity of the waters of the liquid drainages of the mine in both passive miners and the corresponding residual investigation objects, with particular emphasis on those that were formed at the exit of the tunnel. The primary rocks were altered by intense processes SDM (M-2), because they were the samples with the lowest values in pH that were obtained. They should influence, in a decisive way, the increase in the polluting load of the current liquid drainages that drain toward the main receiving aquifers: the rivers "Santa Lucia" and "The Palm", their respective effluents, like those confirmed in this manner for

Fig. 34.1 Microphotograph of the no-altered pyrite (*left*). Microphotograph MEB of the process final of oxidation of the pyrite (*right*)

which the results were obtained by Cañete et al. (2008), Alfonso et al. (1995) and Figueredo et al. (1998).

Another aspect to consider in the investigation is the high electric conductivity values of the samples: 4 and 9, which are justified due to the presence of the ions of Fe^{2+}, Na^{+}, H_30^{+} and ${SO_4}^{2-}$ dissolved in the residuals and silts, those that denote the falling order of contamination in the areas of the swamps, Santa Lucia's tunnel, and the primary ores where they accumulate, respectively.

With the objective of analyzing, the sequence of reactions that SDM associated with the sulphurous minerals lapses during the processes, several microphotographs MEB-EDAX were carried out for which the analysis was divided into two parts: the first, regarding the transformations that experienced the glasses of the pyrite (cubic morphology), and the second which is explained with a more general perspective to recognize the possible rusty phases of the primary sulphurs and their products of oxidation to sulphates.

In the microphotograph of Fig. 34.1 a curious individual glass of pyrite is observed where it is evident that it occurred during the alteration process SDM (oxidation in presence of micro-organisms), the one that is deformed in its cubic morphology, with very fine needles being observed in its interior that show its possible transformation to sulphates of Fe II according to the chemical reactions of oxidation. The above-mentioned confirms the increment of the contents of CO_2 – organic – that associate to Ferro oxidants bacteria, the ZnO, SO_3, Al_2O_3 and the decrease of the tenor of Fe_2O_3, respectively.

Considering the investigations carried out by Henning and Stöör (1986), the ores sulphurous pyrite, which meteorited the typical morphologies, are identified in fine needles or fiber of iron sulphates II moisturized of the series type *zsomolnokita-rozenita-copiapita*. The above-mentioned was proven in the studies carried out by RRD in the analyzed sample M-1donde, one of these phases in which it was detected, the *szomolnokita*.

34.4 Conclusion

In the samples of residual liquids of the streams and river waters from the passive miners analyzed from "Santa Lucia" and "Castellano", very low values of pH were detected that favored the presence of *S, Zn, Pb and Ba* (mg/L) above the maximum concentration that is established for the environmental norms of water.

References

Alfonso et al (1995) Characterization and treatment of solid effluents and liquids generated by the Complex "I Pray Castellano" and "Santa Lucia". Technical Report, CIPIMM

Cañete C, Day A, Ponce N (2008) Study of the environmental degradation of the mining in the region of "Santa Lucia" in the West of Cuba. Report 5668, National Office of Mineral Reacidces (ONRM), MINBAS, Cuba

Cuban Norm NC-521 (2007) Classification of the residual waters

Delgado Diez B et al (2009) Passive environmental miners' identification and environmental remediation of the acid drainage of the Mine Santa Lucia. Inform of investigation, primer stage, Project 62020, CIPIMM, MINBAS

Figueredo FM (1998) Characterization and treatment of the solid waste and liquids of the sulfometales plant. Measures of environmental mitigation. Thesis of Master, Ability of Chemical Engineering, ISPJAE

Guides for the Environmental Handling of the Acid Drainage of Mine in the Peru (2004)

Henning KH, Stöör M (1986) Electron micrographs (TEM, SEM) of clays and clay minerals. Akademie-Verlag, Berlin, p 350p

Karavaiko GI, Kuznetsov SI, Golomzik AI (1972) Role of micro organism in leaching of metals from prays, Nauka, Moscow, 248 pNorm ASTM D-1293 Determination of the pH value in waters and silts

Schwartz H (1965) Maps of Ibero American of atmospheric corrosiveness, I Project MICAT, XV.1 CYTED Werkst. Korros, 16: 93–103, 208–212. In: Corrosion and protection of metals in the atmospheres of Ibero Americans, leave 1 (1998)

Schwartz H (1972) Maps of Iberoamericans of atmospheric corrosiveness, I Project MICAT, XV.1/CYTED, Werkst. Korros, 23: 648. In: Corrosion and protection of metals in the atmospheres of Iberoamerican, leave 1 (1998)

Silverman MP, Lundgren DC (1959) J Bacteriol 77:642

Zamora G (2006) Physical-chemical treatment of acid drainages of mine: I manage environmental of muds of high final density-stability-disposition or applications (in) clean technologies in the extractive industries miner-metallurgy and oil company, Ibero-American Days CFCE, Santa Cruz of the Sierra, Bolivia)

Chapter 35
Seawater Intrusion in the Coastal Aquifer of Guanahacabibes, Pinar del Río, Cuba

R. Hernández, R. Ramírez, M. López-Portilla, P. González, I. Antigüedad, and S. Díaz

Abstract A considerable part of the aquifers in the world are located in hydrographical open basins, where a contact of the fresh ground water co-exists with the sea water, the latent danger of a marine intrusion (contamination of the aquifer with saline water of the sea) being present in the case of an excessive exploitation of the aquifer or natural phenomenon. The phenomenon of the mixture dilutes sea sweet-water, common in the coastal karst aquifer, acquaintance as marine intrusion, is intensified in a natural way, according to the rain regime and the conditions of recharge of the aquifer or for man's action. The loss of the quality of the natural waters in these aquifers influences negatively in this phenomenon. Many of these aquifers are used to provide water for human consumption and agriculture, as it happens in the coastal area of the peninsula of Guanahacabibes, which provides knowledge of the process of marine intrusion and assures the use of the hydraulic resource of this aquifer. The present work had as its objectives: analyzing the advance of the marine intrusion in the region, monitoring the polluting factors and applying the geophysical methods of hydrochemical modelling, and studying the marine intrusion to appreciate the behaviour of this phenomenon in Guanahacabibes.

Keywords Sea water intrusion • Ecosystems • Coastal karst aquifer • Water pollution • Cuba

---

R. Hernández (✉) • R. Ramírez
Universidad de Pinar del Río, Calle Marti 270 Final, Pinar del Río, Cuba 20100
e-mail: rebeca@geo.upr.edu.cu

M. López-Portilla
Instituto Nacional de Recursos Hidráulicos, Km 1 ½ carretera a Luis Lazo, Pinar del Río, Cuba,

P. González
Chemistry Faculty, Havana University, Zapata between G and Carlitos Aguirre, Vedado, PC 10400 Havana
e-mail: patricia_gonzalez@fq.uh.cu

I. Antigüedad • S. Díaz
Departamento de Geodinámica, Universidad del País Vasco -Euskal Herriko, Unibertsitatea UPV-EHU

H. Farfán González et al. (eds.), *Management of Water Resources in Protected Areas*, Environmental Earth Sciences, DOI 10.1007/978-3-642-16330-2_35,

35.1 Introduction

The peninsula of Guanahacabibes is part of Guane catchment area, and is one of the regions with major ecological importance in Cuba. This flora and fauna richness is possible due to the existence of different ecosystems, which are at the same time a consequence of a complex equilibrium of factors such as water among others.

Natural water in the region obtains its chemical composition by means of diverse processes. The different ecosystems existing in the area present a group of features closely related to the quantity and quality of water; besides there is a co-existence among zones which are permanently or temporarily flooded, together with other arid zones in a variety of different hydrological situations, which needed to be monitored to follow up the marine intrusion in this zone. In coastal and caustic aquifers, the processes accompanying the mixture of fresh and seawater are intensified both naturally and by men action.

The hydrogeological basin of Guane with a surface area of 1,253 km^2 can be divided into three sub-zones according to their geomorphological characteristics: mountain zone of Sierra de los Organos, South Plato and the emerged Peninsula de Guanahacabibes.

This research gives continuity to other investigations carried out by the University of Pinar del Río and the National Institute of Hydraulic Resources in the region. The first stage was devoted to the characterization and searching for the depth of the ground-water table in the geological context and the use of wells to measure the complex system of interaction of water and rocks to complement them with electric conductivity measuring that gives the quantity of mineral diluted in water. Antiguedad (1986) addressed the chemistry of water as a vector of information used as a tool for the indirect study of karst hydrodynamics. The interest lies on the study of the temporal evolution of water chemistry, giving special attention to the representativeness of sampling, and what it means in all hydrochemical stages.

On the other hand, much interest was paid to rain water chemistry in the region because this water supplies the entering chemical information of the water systems and constituents, therefore, an important reference to understand the chemistry of out-flowing water and for the evolution of karsts in the Peninsula de Guanahacabibes. The geophysical surveys contributed to the identification and understanding of the subsoil structure and the nature of its constituents to corroborate the results and achieve better conclusions and recommendations.

35.2 Study Area

Hydrologically speaking, the unique recharge to consider for this aquifer is from a precipitation source, and from the lagoon complex existing in the area. The mean annual precipitation rate in this basin is 1,435.6 mm/y and the evaporation rate from the water layer is 1,512 mm. Mean annual temperature varies from 22.7 °C and

27.5 °C registering a maximum of 34.6 °C and 6 °C (Source INRH). The groundwater flow has a predominant direction to the sea. López Portilla (1999) states that in the testing spots numbered by 1, 2, and 12, the inversion of gradients occurred. It can be explained firstly by the fact that the water table is very variable and depends upon the values of diffusivity and extends both depending upon the aquifer lithology. The hydraulic gradients are approximately from $1*10^{-5}$ to $3*10^{-3}$. These higher values belong to the southern part of the aquifer. This hydrodynamic change may have been produced by marine tide effects.

Sediments and rocks that are part of this complex aquifer of the studied basin are the main features of the upper part of a Great Cartesian Basin, named Great Basin of the Northwest Caribbean, according to Hernandez (2000), with its major development over the waters of the Caribbean Sea. The aquifer is formed by carbonaceous and stratified rocks. The basement of this artesian basin is represented by Jurassic formation (San Cayetano and Guasasa fms.). As the characteristic in the study sector, the aquifer strata shows phreatic sediments (without pressure), where local pressures may occur particularly in those zones where sub-laying clay lithology is dominant. In deeper structures semi-confined and confined strata (with pressure) exist, well structured by its impermeable floor and roof (clay; Hernandez 2000).

- Marine deposits-Quaternary materials, formed by sandy-clay sediments, clay-sand and clay. The approximate width is 10–15 m.
- Alluvial deposits-Early Quaternary materials, formed by sand and gravel materials, sandy clay and clay. The width is very variable from 0.5 to 9 m.
- Alluvial and pro-alluvial Deposits-Late Quaternary materials, formed by sand and sandy clay, clay sand and clay composition.
- Neogene-Quaternary Deposits-Sandy materials, sandy-clay and clay materials.

From the lithological point of view, the complexity of aquifer strata is due to the vulnerability, width and grain size of the intercalations of gravel and sand that are very variable. The width varies from a few centimeters and some meters and the grain size varies widely from very fine sands (0.1–1 mm) to gravel (5–10 mm).

Among the carbonaceous rocks conforming the main collector of the aquifer are found the organogenous limestone proper of the Jaimatitas and Vedado Formations from the Quaternary age that are upper geological layers of the zone closer to the coast and Paso Real Fm, from the Miocene age that occupies the lower part of the layer.

The feeding zone located at the north sector of the study area is represented by:

1. Alturas de Pizarras del Norte (J_{1-2}): covering almost all mountain zones. It is formed by terrestrial rocks from San Cayetano fm., covered by small layers of clay sands.
2. Mogotes zone (J_3): covers the central part of the mountain zone. It is formed by carbonaceous deposits from Jurassic superior and in part from terrestrial deposits from the Jurassic Inferior-Medium.

The transit and storage zone is represented as follow:

1. Plato formed by deposits Neogenic-Quaternary: Occupies the northern territory of the Southern Plato, formed by sandy clay deposits.

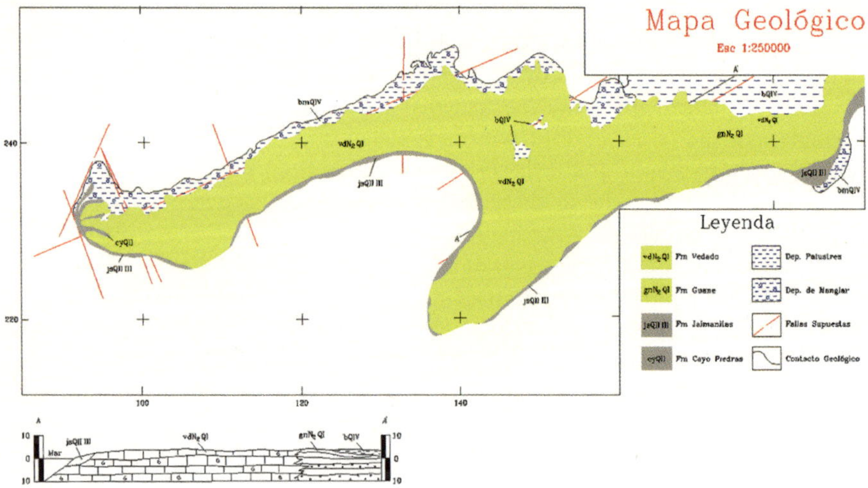

Fig. 35.1 Geologic map

2. Alluvial-pro-alluvial Plato: occupies the wide accumulative marine plato, formed by sandy clay deposits from the Quaternary period.
3. Costal swampy Plato: occupies coast and swamps territories and some parts of the Cuyaguateje, Guadiana, Verde rivers mouth and the Remates Swamp. It is formed by layers of limo, sands, and clays.
4. Karstic and accumulative Plato: Encompasses the abrasive-accumulative of the Peninsula de Gauanahacabibes, situated to the South of Los Remates Swamp. There is an important layer of sandy clay sediments.
5. Abrasive-Karstic Plato: Encompasses the majority of the territory of Peninsula de Guanahacabibes. It is formed by cavernous limestone from Quaternary.
6. Swampy-costal Plato: Occupied mainly the north coast of Peninsula de Guanahacabibes, as a swamp belt of about 1.5–4 km long.

The tectonic of the area is not complicated, which is formed by sediments from the Oligocene López-Portilla (1999). According to this researcher, the region has been affected by light vertical movements. This may clarify the dominance of horizontal stratification in the zone; the pre-Oligocene structure is complex, characterized by the presence of over sliding separated by disjunctive structures, great isoclinal folds and a high degree of fissuring.

The Guane Basin has three geomorphologic sub-zones: Sierra de los Organos sub-zone, Southern Plato sub-zone and the emerged sub-zone of Peninsula de Gauanahacabibes. The mountain sub-zone of Sierra de los Organos represented by Cordillera de Gauniguanico is cut by the Pinar Fault. It is possible to separate the Peninsula as a territory which encompasses from Cabo de San Antonio to the contact limit of sediments from Neogenic and the Jurassic, emphasising: Karstic abrasive Plato, Accumulative Karstic Plato and Swamp Coastal Plato (Fig. 35.1).

In this territory the following geological formations emerge according to Alvarez (1991): Guane, Vedado, Jaimanitas and Cayo Piedra, all of them of very young age, coming from the Neogenic-Quaternary. The lagoon and mangrove deposits are located along the north coast belt and in some inner spots as the case of Los Remates Swamp, which controls the regime of ground-water flow in the northeastern part of the aquifer. Here, a partial discharge occurs from the aquifer as was said by Hernandez (1997). From the lithological point of view there is a homogeneous relationship among the aquifers in the region, the carbonaceous one without much difference, what has not happened in Guane fm., which has different characteristics according to the lithology and impermeability of the aquifer.

35.3 Materials and Methods

The main objective was the analysis of marine intrusion and the position of the salt water wedge at Guane aquifer, using physical hydrology techniques and confirmed with geophysical methods and the PhreeqC 2.80 hydrochemical modeling (Merkel and Planerfriedrich 2002), to integrate the information and analyze the results.

The analysis of the water evolution came from studies carried out in the region by the University of Pinar del Rio and the Institute of Water Resources with the collaboration of the University del Pais Vasco, and University of Mining and Technology from Germany. During the fieldwork the following variables were measured: temperature, pH, electrical conductivity, Redox potential and dissolved oxygen using portable equipment and the content of CO_2 and H_2S by means of standard analytical techniques were found, while micro-organisms were determined by ICP-MASAS.

35.4 Interpretation of Hydrological Results of Guane Basin

The analysis of electrical conductivity measures presented different behaviors, in such a way that low levels of conductivity in areas closer to the swamp exist and the well GB-115, far from the coastal zone. Likewise, some measures of electrical conductivity that were taken about 8 km from the coast show an increase in electrical conductivity which is a consequence of tide penetrations and the mixture of sea water and the fresh water from the swamp. Nevertheless, many artesian wells reported high levels of electric conductivity, which may be a consequence of natural contamination of water.

In some water wells where high levels of conductivity were measured, ranging from 1,000 to 3.68 mS/cm as a consequence of anthropogenic contamination and the existing seawater penetration, as well as the presence of clay materials that may contribute to a higher mineralization and therefore higher dilution. At the Peninsula, measures taken in the central part as well those taken in the shoreline showed lower values of conductivity (803–1,340 μS/cm), which can be due to the presence of sandy material in the area producing low mineralization and by the fact that these wells are located very close to the discharge zone where water infiltrates rapidly.

Table 35.1 Analytical error calculations in PhreeqC

Water well	Depth (m)	% error
115	13	18.02
CAM	5.5	3.68
UQ	4	18.65
CC	5	12.96
CC	10	8.27
CC	12	4.26
18	5	1.88
18	35	−4.97
18	40	2.26
33	2	14.30
33	20	12.66
33	30	−6.60
33	35	−3.87
33	38	−2.75
Sea water	0	−6.32

Table 35.2 Inverse modeling results for the well 33

Depth (m)	Model number	Seawater (%)	Fresh water (%)
30	1	20.73	79.31
35	1	38.71	61.29
38	1	50.31	40.19

The modelling of the proportions of salt with respect to fresh waters carried out by PhreeqC with relation to the salt-fresh water equilibrium and the analytical errors are shown in Table 35.1 (Falb 2004)

According to the obtained results shown in Table 35.1, the inverse model can be calculated to the wells CC with a depth of 10–12 m, as well as to those with 18–5, 35 and 40 m deep and lastly to well 33 at a depth of 30, 35 and 38 m, because the percent of error is less than 10 %. Seawater is type Cl-Na (Na^-: 0.486 mol/L y Cl^-0.573 mol/L). The ionic strength is 0.6924 mol/L, which shows high mineralization. The electric balance (eq) is −0.07755 and the percent of error is −6.32 % (Table 35.2).

Well 33 is an example to explain the characteristic of the mixture between fresh and sea water in the vertical section and the variation between a 21 % and 51 % of the mixture of those waters in 8 m.

The results in well 18 are evidence of a possible fissure at 35 m deep, that at the sampling moment there was one with no equilibrium, what might be indicating that is of recent origin (Table 35.3).

Well CC shows the lowest relation of sea water with respect to fresh water from the aquifer; the interface of sea water and fresh water should be below 12 m deep, although a zone might exist with mixtures with sea water and fresh water over 10 m (Table 35.4). The vertical electric sounding is shown in Figs. 35.2 and 35.3.

Table 35.3 Inverse modeling results for the well 18

Depth (m)	Model number	Seawater (%)	Freshwater (%)
5	2	3.2–3.6	96.9–96.4
35	2	57.8–53.7	46.2–46.3
40	2	2.504	97.50

Table 35.4 Inverse modelling results for the well CC

Depth (m)	Model number	Seawater (%)	Fresh water (%)
10	2	4.04–4.14	95.97–95.87
12	2	1.19–0	98.06–100

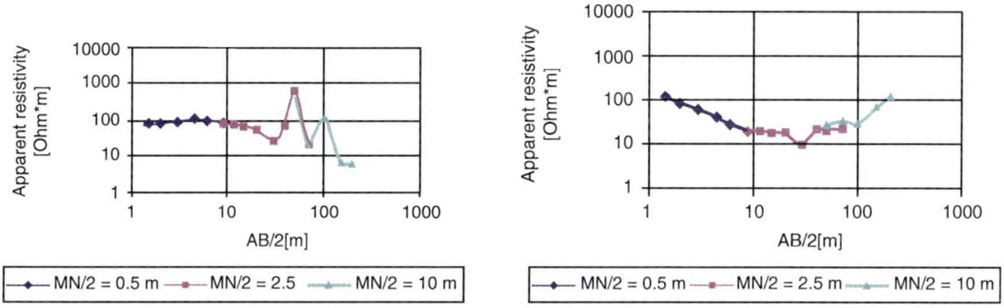

Fig. 35.2 SEV of wells 33 y 18

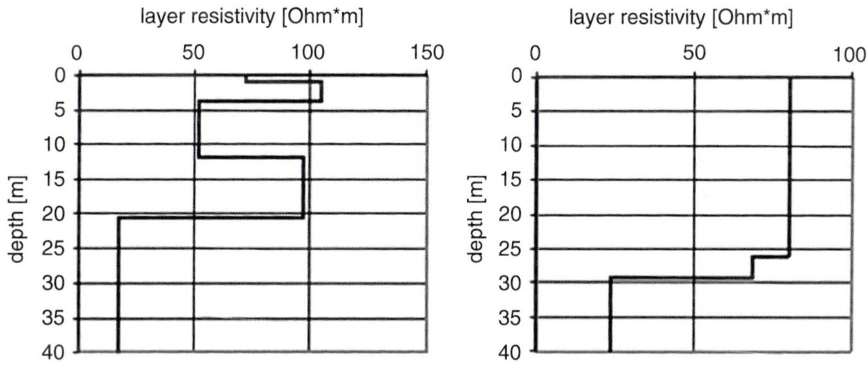

Fig. 35.3 Modeling of geological layer of well 33

With the geophysical survey, the existence of different types of curve in the area was proved. This means that resistivity values in the vertical section are different, which shows the different behavior of marine intrusion in the region. General trends of the curves show one form of H for SEV-18 and a form Q for SEV-115 (H: $\rho 1 > \rho 2 < \rho_3$, SEV −33); the type of curve responds to HKHK ($\rho_1 > \rho_2 < \rho_3 > \rho_4 < \rho_5 > \rho_6$).

35.5 Conclusions

1. The sector studied represents a semi-confined behaviour in the swamp zone and confined in the strata of major depth. The hydrological behaviour in this unit influences decisively on the transmission of saline pollution that is caused by changes of the aquifer permeability. The higher permeability in the aquifer, the higher transmission of saline pollution.
2. The research allowed the evaluation of the actual conditions of the aquifer and set limits on those sectors already contaminated by saline intrusion and for those in a process of salinization in the near future. Pasada de Marin Sector and Las Martinas located to the south is where salinization is well developed.
3. In Uvero Quemado, La Bajada and Cabo de San Antonio the aquifer behaves as open-free to the sea, and the fresh water is situated as a lens over the sea water, producing equilibrium between them that could be disturbed by rain.
4. In those sectors closer to the coast an inversion of flow occurs due to the effect of the sea tides.

References

Álvarez F (1991) Léxico Estratigráfico de Cuba. Biblioteca del Instituto Nacional de Geología y Paleontología. La Habana, Cuba, 672pp

Antigüedad I (1986) Estudio hidrogeológico de la cuenca del Nervión –Ibaizabal: Tesis doctoral, 338pp

Falb J (2004) Seawater intrusion into the neogenic-quarternary aquifer of Guanahacabibes, Pinar del Río, Cuba, 61pp

Hernández R (1997) Caracterización Hidrodinámica e Hidroquímica del Acuífero Neógeno-Cuaternario de Cuenca Guane. Tesis de Maestría ISPJE, 100pp

Hernández R (2000) Caracterización de la salinidad del acuífero Neógeno-Cuaternario de Guane. Tesis Doctoral, 250pp

López P, Guerra P, González B (1999) Investigaciones hidrogeológicas realizada para el abasto de agua al poblado de la Bajada península de Guanahacabibes, 36pp

Merkel BJ, Planer-Friedrich B (2002) Grundwasserchemie- Praxixorientierter Leitfaden zur numerischen Modellierung von Beschaffeneit, Kontamination und Sanierung aquatischer Systeme, Springer, Berlin, 219pp

Chapter 36
Seasonal Dynamic of the Vegetation at "Los Carneros" Lagoon: Handled Floral Reserve Sabanalamar-San Ubaldo, Pinar del Río, Cuba

V. Pérez

Abstract In this study, the seasonal dynamic of the aquatic and swampy plants above ecosystems of sand of quartz is analyzed. The study area is located in the phytogeographical district of Sabaloense, and is characterized by quaternary geology (Pleistocene), with deposits of quartz sands, as well as with pronounced acidic to very acidic features, and is extremely poor in nutrients. The methodology is based in fieldwork in the area from 2005 to 2009, location of three 1 × 20 m large plots for each body of water, with a strict observation of the environmental conditions, and species counting. Through the utilization of the abundance-dominance curves, applications of Shanon and Simpson indices, and by the testing methods of Kruskall Wallis, significant differences among the categories within the inventory, habitat and biological types were determined. Geoelements were also determined, and vegetation profiles were also done at a scale of 1:25 to attempt to represent the natural way in which the species are excluded in the different periods. The results exhibit four states for the period of study, and proven as well was the constant danger to which they are subjected by the wrongful use of these bodies of water.

Keywords Aquatic plants • Geoelements • Ecosystems • Cuba

36.1 Introduction

Cuba, an oceanic island country, with a well recognized floral abundance of approximately 7,020 species, the island has the greatest number of endemic species approximately 50 %, of vascular categorized plants of which 997 of the species are threatened. On the other hand, among the Cuban provinces, Pinar del Río, the most

V. Pérez (✉)
Botanical Garden of Pinar del Río, Km ½ camino al Hoyo del Guamá, Comunidad La Cabaña, Rpto Celso Maragoto, Pinar del Rïo, Cuba
e-mail: odigen@princesa.pri.sld.cu

H. Farfán González et al. (eds.), *Management of Water Resources in Protected Areas*, Environmental Earth Sciences, DOI 10.1007/978-3-642-16330-2_36,

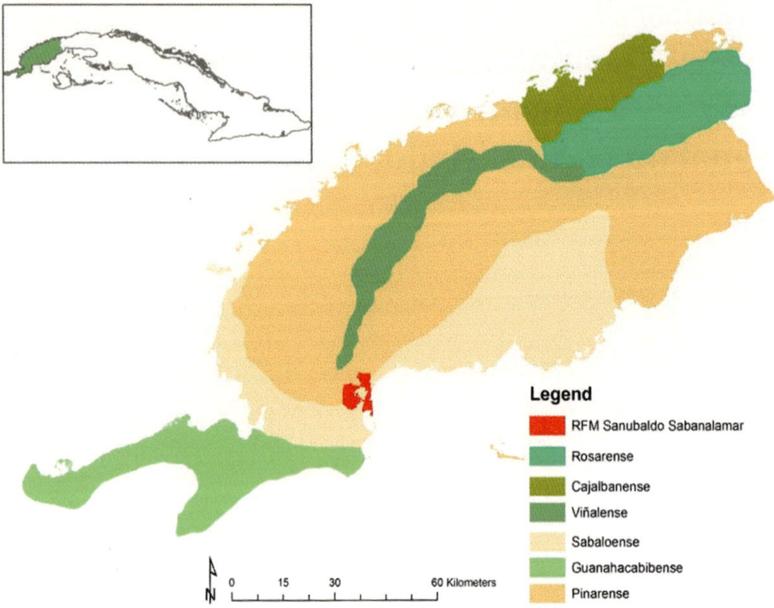

Fig. 36.1 Phytoregionalization de Pinar del Río (Borhidi 1996)

western of all, accounts for 3,278 groups of plants, with about 940 that are endemic. These numbers represent 29 % of the whole, including 346, which are threatened (Urquiola et al. 2010).

The particular case of the aquatic and swampy plants above ecosystems of sand of quartz from the phytogeographical district of Sabaloense (Borhidi 1996), conditioned by substratum and lack of water, suffer from a stern loss of its areas of extension and inhabited place, not only by destruction of its habitats, but also due to the dangerous increase of various types of substances being discarded into the ecosystem, and the changes that are generated in the features of these lagoons.

36.2 Study Area

The political-administrative province of Pinar del Río, extends from the SW to the NE, occupying an extended area of 10,904 km^2. Its western boundary is with the Channel of Yucatan, which separates it by 210 km from the Cape of Catoche in Mexico. To the north are the waters of the Gulf of Mexico, to the south the western Caribbean Sea, and to the east the province of Havana. However, as of January 2011, due to a reorganization of the political-administrative division of the country, the province to the east is the newly formed province of Artemisa. According to

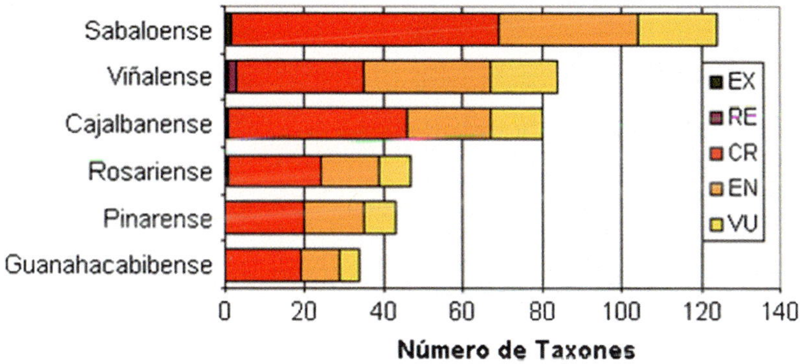

Fig. 36.2 Comparative chart between taxones and districts

Borhidi (1996), the province of Pinar del Río, is divided into six (6) phytogeographical districts (Fig. 36.1)

The "Sabaolense" district extends for the low plains northwest of the Mantua municipality, through Guane, Sandino, San Juan, Martinez, San Luís, Pinar del Río, Consolación del Sur and Los Palacios. In the Sandino municipality the district extends to the north of the "Los Remates" swamp (The Black Ones), unto the Grifa and Cortés.

The work area is characterized by quaternary geology (Pleistocene), with deposits of quartz sands, with pronounced acidic to very acidic features and is extremely poor in nutrients. This sandy quartz substratum counts on the largest number of threatened taxones in the province of Pinar del Río with a total of 123 (Fig. 36.2) in comparison with the other districts (Urquiola et al. 2010). Their annual precipitations range from 1,000 to 1,400 mm with periodic flooding due to layers of clay beneath the sand substratum. The average temperature is 24.6 °C.

The Los Carneros lagoon is located at 22° 07′ 00″ N and 84° 00′ 22″ W, situated in the Sabanalamar-San Ubaldo Handled Floral Reserve.

36.3 Methodology

It is included in the fieldwork performed during expeditions to the area from 2005 to 2009, three locations with plots for each body of water measuring 1 × 20 m, each with a strict observation of the environmental conditions, and species counting. The office work started by the bibliographical research about the target area, checking the herbarium collections and the reordering of the outcomes. Abundance-dominance curves were determined through the application of Shanon and Simpson indexes, by means of the test of Kruskall Wallis. The significant differences were tested among categories of inventory, habit and biological kind (Ellemberg and Mueller-Dumbois 1967a, b). Geoelements were also determined (Borhidi 1996), and vegetation profiles were also performed at a scale of 1:25 to attempt to

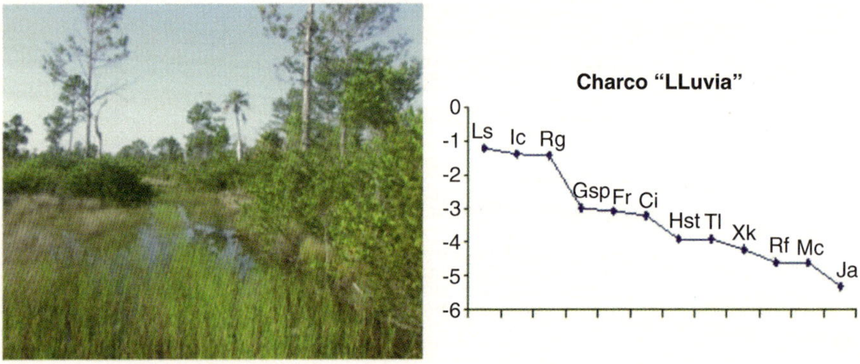

Fig. 36.3 Puddle in the border. Abundance curve

represent the natural way in which the species are excluded in the different periods. Specialists of diverse specialties such as botanist and ecologists were interviewed.

36.4 Results and Discussion

The vegetation in the "Los Carneros" lagoon constantly exhibits a changing annual behavior, proven during 5 years of research, always associated with the water, given by the variations in the quantity of precipitations, like for accumulation of this in the substratum. These marked seasonal dynamics could be focused in four stages:

It was determined by the first stage, which occurred between May to July that the humidity process commences in the substratum when the humid hydrological period begins. This is utilized by various plants such as: *Xyris jupicai, Xyris ekmanii, Xyris elliottii, Uricularia juncea, Utricularia foliosa, Utricularia purpurea, Utricularia resupinata, Rynchospora globosa, Rynchospora fascicularis, Caperonia palustris, Eleocharis cellulosa and Chrisobalanus icaco,* to initialize their development.

The second stage took place between July and the first 2 weeks of November, an interval in which the largest levels of precipitation occur in this fluvio-marine and lacuna-marshy fringe of the south and south occident of Pinar del Río, As was previously explained, with regard to the composition of the soil that is present in these lagoons, then the soil is oversaturated and a flood occurred, exceeding the limits of each, and as a consequence of this, they end up merging with one another along some of their borders. For example, there is a connection between this lagoon and a distant puddle from the border (Fig. 36.3). In this stage, a peculiarity remains with the water, only during the month of August, *Isoetes cubana* is observed. A species that had its focus for several decades classically in the Jovero lagoon. This ecosystem has been subjected on a yearly basis to an intense agricultural, livestock

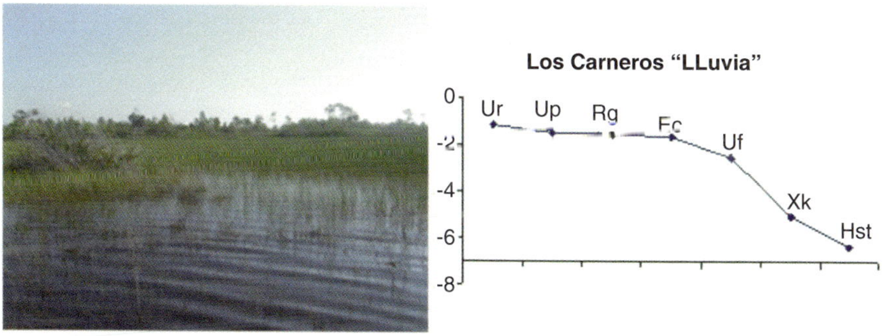

Fig. 36.4 Picture of Los Carneros in raining season. Abundance curve

Fig. 36.5 *Bacopa monnieri* (**a**), *Drosera capillaries* (**b**), *Ludwigia torulosa* (**c**), *Micranthemun rotundatum* (**d**)

and aquatic exploitation, which could not be observed during this investigation beyond the San Felipe’s Plateau in Camagüey, of which there are no current reports.

When this flood occurs, the first juvenile of the rest disappear, *Xyris, Chrisobalanus, Rynchospora y Utricularia,* with the exception of *Utricularia resupinata, Utricularia purpurea y Eleocharis cellulosa* (Fig. 36.4)

For the third stage, that includes the last 2 weeks of the month of November until January, began the drought period, and with it, the lagoon exhibits its maximum splendor with the appearance and flowering of most of the species that inhabit the area. Once again it is observed that *Xyris jupicai, Xyris ekmanii, Xyris elliottii, Uricularia juncea, Utricularia foliosa, Utricularia purpurea, Utricularia*

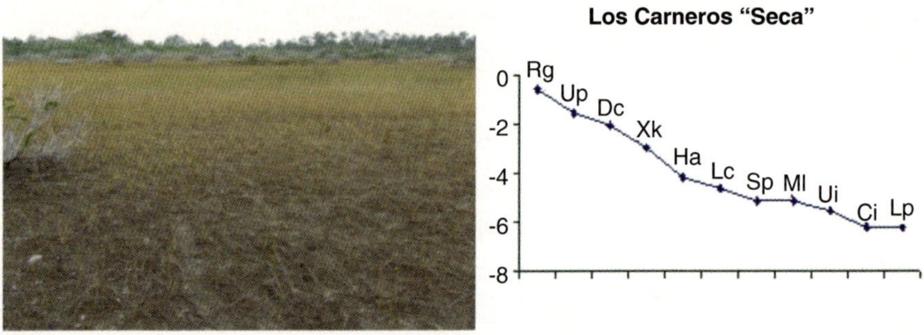

Fig. 36.6 Picture of Los Carneros in drought season. Abundance curve

resupinata, Utricularia pumila, Rynchospora globosa, Rynchospora fascicularis, Caperonia palustris, Hydrolea spinosa, Eleocharis cellulosa and Chrisobalanus icaco, Ludwigia stricta, Ludwigia torulosa, Micranthemum rotundatum, Encopella tenuifolia, Bacopa monnieri, Phyllanthus eliotropus, Phyllanthus aquaticus Eragrostis charis, Drosera capillaris, Burmania bicolor, are in the borders when the benchmarks of the water levels return to their normal limits. Most of them are minute grasses of their habit, and they do not share the same area inside the lagoon. That made it extremely difficult to include them all within one photograph alone (Fig. 36.5).

The fourth stage is characterized by the presence of a second drought period more intense than the previous one that extended from the month of February until April. In this time the lagoon completely dried up, (Fig. 36.6) and the water level is below the substratum at a depth between the 45 and 70 cm. Only some mature copies of perennial species survived, such as *Ludwigia torulosa, Eleocharis cellulosa, Chrisobalanus icaco, Hydrolea spinosa, Acoelorraphe wrightii, Cyperus giganteus e Indigofera mineata.*

36.5 Conclusions

Through these results, the marked presence or absence of the existent aquatic and marshy species was proven to be in the "Los Carneros lagoon". Also proven was the the constant danger to which these species are subjected due to the wrongful use of these bodies of water and exploitation of their areas by means of agricultural, livestock and aquatic purposes, as well as extensive raising of cattle. These activities continue to contribute organic matter to the substratum. The continued dumping of solid substances and the loss of biodiversity in these acidic lagoons increases the risk of change to the chemical features within the Sabaloense district.

References

Borhidi A (1996) Phytogeography and vegetation ecology of Cuba. Akademiai Kiado, Budapest, 858 pp

Ellemberg H, Mueller Dumbois D (1967a) A key to Raunkiaer plant life forms with revised subdivisions. Borg. Geobot. Inst. Rübel

Ellemberg H, Mueller-Dumbois D (1967b) Tentative physionomic ecological classification of plant formations of earth. Borg. Geo. Inst. ETH, Stoftg. Rübel, Zürich

Urquiola AJ, González-Oliva L, Novo R (2010) Libro Rojo de la Flora Vascular de la provincia Pinar del Río

Part VI
Information, Popularization and Training

Chapter 37
Toward the Sustainable Use of Water in Cuba

J.A. Díaz Duque

Abstract The main problems related with the sustainable use of water in Cuba are: the low volume of available water for inhabitants per year; a water stress near 50 %, an indicator of the shortage of water of 50 %; a water footprint that surpasses the 1,700 cubic meters for inhabitant per year; a low index of annual replacement of water resources with 13.7 %; the increment of the flow of residual waters and their low use; the drop efficiency in the use of water and the considerable losses in the distribution and consumption nets. The National System of Protected Areas has guided its strategic projections toward the protection of the natural resources of the country, with special emphasis on the biological diversity, for which the water resources have been the center of attention. Water constitutes for Cuba the main environmental challenge to guarantee its development, as well as its environmental and alimentary security, which will only be possible by means of sustainable management of its water resources, on the basis of efficiency, saving, and its protection.

Keywords Management of water resources • Sustainable water resources • Protected areas • Cuba

37.1 Introduction

The total volume of water in the Earth is approximately 1,400 millions of km^3 of which only 2.5 %, that is to say, around 35 million km^3, correspond to fresh water for which a larger volume accumulates in the poles, in Greenland and in very deep deposits. Only about 0.01 % of all the water of the planet, approximately 200,000 km^3, is profitable for human use, when coming from lakes, rivers, wetlands

J.A. Díaz Duque (✉)
Departamento de Geociencias, Instituto Superior Politécnico "José A. Echeverría", CUJAE, Calle 114, No 11901, e/119 y 127, Marianao 15, Ciudad de La Habana, Cuba
e-mail: jaduque@civil.cujae.edu.cu

H. Farfán González et al. (eds.), *Management of Water Resources in Protected Areas*, Environmental Earth Sciences, DOI 10.1007/978-3-642-16330-2_37,

and shallow underground basins (UNEP 2003). On the other hand, the distribution of this volume of water is very heterogeneous and unequal, existing in regions and countries with great abundance and others with very scarce water resources.

Experts have defined the threshold of water stress (Falkenmark and Widstrand 1993). Indicating it as 1,700 m^3/inhab/year, it calculates starting from difficulties that do not exist for the development of human activities, as long as the values are inferior for the mean shortage of the resource it dilutes that they imply serious limitations for sustainability, particularly when this indicator is located below 555 m^3/inhabitant/year. From the appearance of human beings, the quantity of water per inhabitant per year has diminished, particularly in the last 200 years. For example, between 1950 and 2000, the availability of water per person diminished from 16,800 to 6,800 m^3 a year (Shiklomanov 1997). The world consumption of water grew twice during the last century to a rate superior to that of the population. From 1950, the world surface dedicated to the irrigable one was duplicated, and the use of water for agriculture, industry and domestic consumption tripled (UNESCO 2006).

When concluding the process of the formation of the Earth and the accumulation on its surface of enormous volumes of water, approximately 4,000 million years ago, the formation of life began, first with the appearance of the simplest organisms and opening the way to more complex forms, as much in plants as in animals. Finally, the Homo sapiens and their expansion in the whole planet demanded larger volumes of water to guarantee their feeding and conditions of life: agriculture, industry, constructions, transport, and recreation. Gradually it decreased the physical vital area of the planet for each inhabitant and with it the readiness of natural resources, in particular, water resources that, in turn, as a consequence of ¨civilization¨ would be affected by the constant increment of contamination (PNUD 2006).

37.2 The Situation of the Water Resources in Cuba

The fresh water potential of Cuba ascends to 38.1 km^3, of which 31.6 km^3 (73.4 %) corresponds to the superficial waters and 6.5 km^3 (26.6 %) to the underground waters. Of this potential, 24 km^3 are profitable and available, 13.6 km^3 with a bigger incidence in the superficial waters (67 %; Cuban Ministry Of Science, Technology and Environment 2008). The hydraulic development of Cuba has facilitated to use 57 % of the profitable water resources, by means of creation of the technical pertinent infrastructure, which will increase to 200 times the capacity of the reservoir of the country and will provide 96 % of the whole population with access to drinkable water and 95 % to sanitation (Cuban Objetivos de Desarrollo del Milenio 2005; García Hernández 2007).

The Environmental National Strategy 2007–2010 (ENS) identified for first time the lack of water as one of the main environmental problems of Cuba, even after the Cuban development hydraulic had elevated the reservoir capacities to more than 9,600 million cubic meters from 1959 (Cuban Ministry Of Science, Technology and

Environment 2008). In such a sense ENS expressed: "... *water shortage remains high for meeting economic, social, and environmental needs on the island. The fact is aggravated by natural disasters (prolonged droughts and changes in seasonal patterns) as well as man-made impacts, (including saline intrusion, overexploitation, pollution, etcetera)*".

Among the factors of natural origin that intervene in the lack of water in Cuba they are distinguished as:

1. The insular character of the country conformed by a big lengthened and narrow island and numerous islands, keys and islets.
2. The disposition and rugged structure which determines the existence of a water power station throughout the largest island in the direction of their longitudinal axis, with two slopes, northern and southern, in those that run almost all the rivers, in traverse address, with changes in their flows in dependence on precipitations and with a relatively short course.
3. The extraordinary extension of the karstic phenomenon. Almost 70 % of the areas that occupy the aquifer formations are of karstic origin.
4. The prevalence of rivers of small longitude (smaller than 40 km) and hydrological basins with less than 200 km^2 of surface.
5. The location of the Cuban main aquifers in the coastal areas, with an opened character for which their waters are in direct contact with the seawater.
6. The drop natural readiness of the water, the only one that reaches 1,221.5 m^3 for inhabitant per year for all the uses.
7. The falling tendency of the historical national recording of precipitations, when diminishing 133 mm in the period 1961–2000 with regard to the previous period 1931–1960, being the new recording of reference of 1,335 mm. Falling in the three regions of the country, particularly in the oriental region with 260 mm less, with negative singular impact in the basin of the Cauto River with 367 mm less and in the Guantanamo-Guaso basin with 154 mm less (Cuban Servicio Hidrológico Nacional 2006).

The impacts of climatic change are manifested in the hydrological regime and in the readiness of water, particularly for Cuba for their archipelago condition. The increment of the mean temperature in the Earth harnesses the elevation of the level of the oceans and seas for which the saline wedge will advance toward the interior, increasing the salinization of the coastal aquifers and therefore diminishing the volume of profitable fresh water. The processes of drought will take place with bigger intensity, territorial extension and will frequently affect the reservations of fresh water, as much superficial as underground.

37.3 The Water Resources in the Protected Areas of Cuba

The National System of Protected Areas (SNAP) of Cuba has been conceived with a total of 263 spaces, of which 80 have national significance, including the well-conserved ecosystems with the natural largest values in the country. The system in its group embraces 22 % of the national territory SNAP, which has international important recognitions: six Biosphere Reserves, three World Patrimonies and six Ramsar places (Cuban Ministerio de Ciencia, Tecnología y Medio Ambiente 2009a).

The main headwaters of Cuba, which give origin to most of the hydrological basins of national interest, take place in the mountains, ecosystems that are of singular importance in the formation of the water resources of the country: Guaniguanico, Bamburanao, Guamuhaya, Sierra Maestra and Nipe-Sagua-Baracoa. The total area of these mountains represents 18 % of the surface of the country and in them inhabits 21.5 % of Cuba's population Cuban Oficina Nacional de Estadísticas (2008).

SNAP of Cuba has guided its strategic projections toward the protection of the natural resources of the country, with special emphasis on the biological diversity, for which the water resources have been in the center of attention, including some of the legally established protection categories, the headwaters and part of the course of the main rivers, specific areas of basins of national interest and all the mountainous ecosystems.

The pollution of the water bodies by reason of the dumping of residual wastes of the human establishments, the coffee pulp removers and the swinish farms constitutes the main problem in the sustainable handling of the water resources in the protected areas, with a strong impact toward the inferior areas of the hydrological basins near the outlet of the rivers, in which some of the phenomenon of salinity is also manifested.

37.4 Main Problems Related to the Sustainable use of the Water Resources in Cuba

The executive summary of the report GEO CUBA (2007), when considering the state of the resource dilute, points out: *¨is considerable the losses of water in the conduction process and distribution due to the not well state of the technical nets and to the use of inadequate technologies, fundamentally in the watering of the agricultural cultivations. It should be increased, therefore, the efficiency in the use of the water and the water and hydraulic environmental culture to mitigate the effects of their relative shortage, in a climatic complex context (persistent droughts and hurricanes), all that which elevates the transmission of illnesses of water origin like the sharp diarrheas illnesses (EDA) and the hepatitis¨* (Cuban Ministerio de Ciencia, Tecnología y Medio Ambiente 2009b).

The full text of the report (Cuban Ministerio de Ciencia, Tecnología y Medio Ambiente 2009c) indicates:

- The flow of residual waters that enters the natural waters surpasses the capacity of self-purification of these last ones.
- In the hydrological basins of national interest were disposed an estimated pollution load of 25,538 Tm (BOD_5), emitted by a total of 495 punctual sources.

- With regard to the deterioration of the quality of the water, studies of the last three decades (1970–2000) reflect the displacement of the wedge of marine intrusion in the aquifers in direct contact with the sea which in a horizontal sense, has reached values between 0.3 and 3 km/year inland; and in the vertical one the mixture area among the fresh water and salted has ascended between 0.5 and 5 m/year.

There are several conventional indicators to assess the sustainability in the use of water resources:

- The availability of water resources expresses the volume of available water for inhabitant a year for all the uses. Below the 1,700 m^3/inhabitant/year it indicates a shortage of water.
- The water stress or appraises among extraction and readiness measures the extraction of the population's total annual water and it compares it with the available water resources. It is expressed by percent to show that the larger the rate, the larger the water stress.
- The water footprint relates the total volume of water consumed for the elaboration of goods and services that are consumed by the inhabitants of a country. It includes the goods and services taking place nationally as those cared products. It is expressed in m^3/inhabitant/year.
- The shortage of water indicates the relationship between the consumption of water and the availability of the water resources. It is expressed as a percent of the water footprint regarding the available water resources for inhabitant a year.
- The virtual water points out the necessary volume of water to elaborate a product or to facilitate a service. It is expressed in terms of cubic meters of water by unit of mass or of volume of the product or for unit of the executed service: m^3/kg, m^3/t, m^3/l, m^3/rented room.
- The environmental flow points out the minimum volume of water that should be used by each natural ecosystem to guarantee their survival and operation.
- The index of restock of the water resources expresses the percentage relationship among the extraction volumes and of the natural annual reinstatement of the water resources.
- The storage capacity indicates the volume of water dammed by inhabitant.
- The index of contamination of the waters expresses the rate (%) of the flow of the residual waters that enter the water bodies with regard to their flow.
- The recycling of the residual waters and their consequent use.

Experts of INRH (García Fernández and Cantero Corrales 2008) expose the behavior of some of these indicators for Cuba:

Availability of water resources	1,220 m^3/inhabitant/year
Water stress	51 %
Water footprint	1,712 m^3/inhabitant/year
Shortage of water	50 %

Regarding the virtual water limited reports existing for goods and services taking place in Cuba, significant experiences in the sugar industry, in tourism and the alimentary industry, were particularly useful during the intense drought of the years 2003–2005. However, the data and the available statistical series are insufficient to carry out a deep analysis of the efficiency in the use of water in the productive Cuban sector and the comparison with similar ones in other countries. Neither do they allow the appropriate assessment of the flow of water in the commercial exchange; that is to say, all is flooded it exports for example starting from their employment in the goods and services that leave the country: sugar, rum, tobacco, medications and others. Their indetermination makes imprecise the calculation of the water footprint for Cuba.

The environmental flow is another important indicator that has not been sufficiently studied and certain for the main Cuban hydrological basins, which drive a maladjusted pattern of distribution of the water resources in the country, in that the ecosystem takes the worst part, being, however, the one that sustains the hydrological cycle and the means of life through the goods and services that it reports to the society. Evidences of ecological stress have shown important basins like those of Cauto, Ciénaga de Zapata and Cuyaguateje, in those that the anthrophic activity has altered the ecological balance in detriment of the sustainable medium and long-term development. The established order of priority for the annual plan of assignments of the water (Dirección de Obras Hidráulicas INRH 2006) does not identify the environmental flow and it only considers, in the fifth place, a denominated other category, although it recognizes the importance of the one denominated expense sanitarium for the balance of each reservoir.

The recording of the water resources depends on several factors among those that are half the level of the annual precipitations, the evaporation and the variability of the runoff, as well as of the extraction volumes. It has been considered (Planos Gutiérrez 2008) that the annual volume of rain for Cuba is 149 km^3, which contributes about 30 km^3 of water, being the annual extraction of 7.0 km^3, representing 23 % of the same, near 625 m^3/inhab/year. The annual volume of extraction of water is equal to 18.4 %, 29.2 % and 51.2 % of the total, profitable and available water resources of the country, respectively.

On the other hand, the annual index of water replacement for Cuba (PNUMA 2003) is 13.7 %, lightly superior to the threshold of affectation of the future readiness of the resource (10 % as the maximum limit of the natural replacement), which is considered as moderate stress, that is to say, smaller than 20 %.

The water dammed in Cuba has grown more than 200 times regarding the volume of 1959 as long as only the population has made it less than twice as much. This makes the nominal volume of water dammed by inhabitant to ascend to 858 cubic meters. This growth has allowed a larger use of the water potential of the country, basically for agriculture and the population.

Since 2006, an increment of the polluting organic load prepared in the water bodies took place, showing a larger intensity in 2007 with a total increment of 9.7 % regarding the previous year, with appreciable variations in the territories of Havana (70.5 %), Cienfuegos (32 %), Ciego de Ávila (3.2 %), Guantánamo (3.6 %) and Isla de la Juventud (212.8 %). Several hydrological basins were seriously affected by the direct dumping of the residual from swinish centers, among them the Ariguanabo and Guantánamo-Guaso basins. In the country, 2,060 main polluting sources have been identified, those that to the closing of the year 2007 accumulated the disposition of a polluting estimated load of 155,241 tons of BOD. In the same period the main basins received 25,539 tons of BOD, coming from 495 sources; the basin of the Cauto is an outstanding one with a reception of 11,471 t BOD in the year, 45 % of the total.

Even when nationally it is not considered the disposition of the polluting inorganic load, there are punctual reports of the presence of heavy metals and of other chemical compounds in superficial and underground waters, due to the dumping and wrong handling of industrial residual.

The increment of the sewer system service, the access to drinkable (96 %) water and the sanitation (95 %) in the cities and the urban and rural establishments, without the elevation corresponding to treatment (35 %) and the reuses of residual waters, has provoked an increase in the flow of the polluted waters toward the water bodies.

37.5 Road to the Sustainable Management of the Water Resources in Cuba

Through the decades one has worked laboriously to increase the index of availability of the water resources, and it still continues in the construction of channels, tunnels, and reservoirs, with the objective of water arriving to areas with a natural deficit in the resource, particularly in the oriental region and in the capital of the country.

Another important work that addressed the advancement toward the sustainability of the water resources is the radical and quick decrease of the considerable losses of water in the distribution and consumption nets, those that have been estimated at 800 million m^3 a year. The elimination of the leakages requires considerable investments for the extension (more than 21,000 km of nets) and the conduits that, in general, are not in good condition.

The water efficiency is the most important factor in the increment of the availability of water for the Cuban population. The irrigated agriculture represents the largest volumes of water consumed annually (65 %); however, the systems used are very inefficient. The introduction of the located watering or for leak, using the agro-meteorological information, will facilitate not only to reduce the unnecessary consumption of water and to elevate the agricultural yields, but also to protect the soil from erosion and avoid the contamination of the water bodies. Although with a smaller weight, for the total volume of consumed water, the Cuban industry included tourism, which also has the obligation to reconsider their consumption patterns for produced unit or offered service and come closer to the international best standards.

Similar to the above-mentioned, the saving of water will be performed with similar focus and tasks as that of saving of energy. It is not only to design and to implement valuable and necessary environmental education campaigns, it is necessary to measure and to control the expense of water, to apply upward rates in function of the consumption. The footage, the slipping rates, the thrifty devices, the restrictions in the use of the water, essentially the underground water, are some of the main actions for the sake of saving. Punctual experiences indicate the positive benefits of these measures.

The reduction in the contamination in the bodies of water is a task of first priority. In spite of the efforts carried out, it has not been possible to reduce the flows of residual waters coming from human settlements through the sewer system nets, which are also connected to some industries, laboratories and other entities of producers of goods and services.

Once the water has been used in a given process, be it industrial, agricultural or domestic, it is used again after the treatment required for its specific ends or it is restored in a sure way to the rivers or to recharge the aquifers, alleviating its exhaustion. The watering in the sugar complexes, the watering of gardens and golf fields in tourism, and the re-circulation of the water in the cooling towers are some of the countless examples that can make an improvement.

There are other work lines or actions directed to the sustainable use of the water and they should not be overlooked because of their relative complexity:

- The calculation of the environmental flow for the main hydrological basins and the implementation of their strict execution in the annual balance of the distribution of the available water resources of the country.
- The reduction of the evaporation in the reservoirs.
- The creation of underground reservoirs.
- The calculation of economic benefits of the goods and services it offers the ecosystems, particularly the hydrological basins and the wetlands, to demand payments to those productive entities that are served as the same ones.
- To determine the volumes of virtual water, as much the one that is exported as the one that it is cared, in order to evaluate the commercial flow of virtual water of the country objectively.

37.6 Conclusions

1. At the present time serious limitations exist with the available water resources in Cuba, which are heterogeneously distributed in space and time. These limitations are owing to factors of natural and anthropic origin.
2. The occurrence of the extreme natural hydrometeorological phenomena of the recurring droughts and hurricanes, as well as the adverse impacts of climatic change, has increased the picture of the shortage of water in Cuba.
3. Almost all of the indicators that evaluate the sustainability in the use of water in Cuba show a negative action, evidencing a disproportion between consumption and availability, which outlines a challenge for the management of the water resources as for the maintenance of the ecological integrity of the basins and the answer to the growing demand.

4. The solution to this complex challenge is achieved by means of the implementation of a group of actions directed to reach the sustainable handling of the water in Cuba. These actions have as a basis efficiency, saving and protection of the water resources of the country.
5. The National System of Protected Areas of Cuba can guarantee an integral and sustainable management of the water resources through the Handling and Operative Plans of the areas.

References

Cuban Ministerio de Ciencia, Tecnología y Medio Ambiente (2009a) GEO CUBA 2007: Evaluación del medio ambiente cubano. A. Fernández Márquez A, Pérez de los Reyes R (eds). La Habana, 293p

Cuban Ministerio de Ciencia, Tecnología y Medio Ambiente (2009b) GEO CUBA 2007: Evaluación del medio ambiente cubano. Resumen Ejecutivo. Fernández Márquez A, Pérez de los Reyes R. (eds). La Habana, 28p

Cuban Ministerio de Ciencia, Tecnología y Medio Ambiente (2009c) GEO CUBA 2007: Evaluación del medio ambiente cubano. Fernández Márquez A, Pérez de los Reyes R (eds). La Habana, 293 p

Cuban Ministry Of Science, Technology and Environment (2008) National Environmental Strategy 2007–2010. Havana, Cuba, 71 pp

Cuban Objetivos de Desarrollo del Milenio (2005) Segundo Informe de Cuba. Julio de 2005

Cuban Oficina Nacional de Estadísticas (2008) Información Estadística del Plan Turquino. Dirección de Agropecuario. La Habana, 101p

Cuban Servicio Hidrológico Nacional (2006) Nuevos logros en el estudio de la pluviosidad en Cuba: Mapa Isoyético para el período 1961–2000. Revista Voluntad Hidráulica 98: 2–14

Dirección de Obras Hidráulicas INRH (2006) Demanda – Disponibilidad. Una relación a tener en cuenta. Voluntad Hidráulica 98: 37–42

Falkenmark M, Widstrand C (1993) Population and water resources: a delicate balance. Population Bulletin. Population Reference Bureau, Washington

García Fernández JM, Cantero Corrales L (2008) Indicadores globales para la evaluación del uso sostenible del agua: Caso cubano. Voluntad Hidráulica 100: 12–19

García Hernández JM (2007) Aplicación del enfoque ecosistémico a la gestión integrada de los recursos hídricos. Aproximación al caso cubano. Revista Voluntad Hidráulica 99: 2–17

Planos Gutiérrez EO (2008) El agua, una responsabilidad compartida. III Taller Uso y Cuidado del Agua. La Habana, 23 y 24 de junio de 2008. Sociedad Económica de Amigos del País [CD-ROM]

Programa de las Naciones Unidas para el Desarrollo (PNUD) (2006) Más allá de la escasez: Poder, pobreza y crisis mundial del agua. Informe sobre Desarrollo Humano 2006, 422p

Programa de las Naciones Unidas para el Medio Ambiente (PNUMA) (2003) GEO América Latina y el Caribe. Perspectivas del Medio Ambiente 2003. PNUMA, Oficina Regional para América Latina y el Caribe, D. F. México, 281p

Shiklomanov IA (1997) Valoración de los recursos hídricos y disponibilidad del agua en el mundo. Estocolmo. OMM, PNUMA, IMA, 73p

UNESCO, Programa Mundial de Evaluación de los Recursos Hídricos (2006) El agua, una responsabilidad compartida. 2º Informe de las Naciones Unidas sobre el desarrollo de los recursos hídricos en el mundo. Resume Executive. Available in: http://www.unesco.org/water/wwap/index_es.shtml Accessed 24 Aug 2010

United Nations Environment Programme (UNEP) (2003) GEO: Global Environment Outlook 3. Pasado, presente y futuro. In: CD-ROM GEO Reports: Analyses and Outlooks

Chapter 38
Geodiversity and Hydrological Patrimony in High Mountain Areas. Gredos Range: Béjar and El Barco Massifs, Spain. Inventory and Cataloguing

R. Cruz, J.L. Goy, and Cardeña Zazo

Abstract The Sierra de Gredos mountain range is located to the east of the central plateau and separates the basins of the Rivers Duero and Tajo; the tectonic massifs of Béjar and El Barco are formed at the western end. This mountain chain presents great geological, biological and landscape diversity, acknowledged by the declaration of parts of its territory as Protected Natural Spaces (PNS): Gredos Regional Park, Los Infiernos Nature Reserve, Natura 2000 Space (ZEPA and LIC) and Biosphere Reserve. In the Béjar and El Barco massifs, many of the water-related features provide knowledge and interpretation of their natural and social history and are a source of quality of life for their inhabitants and visitors, thus forming part of the Natural and Cultural Patrimony, which require conservation. In spite of the legal protection regime, there is currently a risk of degradation and/or destruction of the most fragile values. To facilitate its sustainable management, an inventory of the Hydrological Patrimony was drafted, mapped in GIS, with 35 A/L/PGI_H, among which are included: glacial lagoons, waterfalls, rapids, summit peat bogs and a snow well. Educational itineraries are designed for their revaluation, helping promote their conservation through knowledge.

Keywords Protected natural spaces • Hydrological Patrimony • GIS • Spain

---

R. Cruz (✉) • J.L. Goy
Department of Geology, Area External Geodynamics, Faculty of Sciences, University of Salamanca, Plaza de la Merced s/n, 37 008 Salamanca, Spain
e-mail: rqcruz@usal.es; joselgoy@usal.es

C. Zazo
National Museum of Natural Sciences, Geology Section, Calle José Abascal n° 2, Madrid, Spain
e-mail: mcnzc65@mncn.csic.es

H. Farfán González et al. (eds.), *Management of Water Resources in Protected Areas*, Environmental Earth Sciences, DOI 10.1007/978-3-642-16330-2_38,

38.1 Introduction

The Sierra de Gredos belongs to the Central Iberian zone of the Iberian Massif (Geological Unit of the peninsular West) and within it is located in the Central System, a sector reactivated as a tectonic massif during the alpine orogeny. The tectonic massifs of Béjar and El Barco were formed in the Plio-Quaternary as high mountain sectors, with an average height of 2,000 m above sea level and a maximum of 2,430 m, and in them the climatic oscillations of the Quaternary became very important. During the cold periods, the glacial and periglacial morphogenetic systems settled in and their actions gave rise to a set of singular shapes and deposits, representative of those periods in which a large part of the fallen water was stored as ice, forming characteristic alpine surroundings. From the last cold episode, in the Late Glacial, the climate gradually shifted to Mediterranean conditions, which became consolidated in the Holocene.

They currently present a Mediterranean climate with clear zoning, which conditioned the establishment of bioclimatic bands and the development of a *hydric system lodged in and adapted to its abundant fractures,* on which many of its natural and landscape values depend. Humans settled in prehistoric times and lived in harmony with the surroundings.

The rich geodiversity, biodiversity and quality of the landscapes provide data that allow an understanding of their evolution and the meaning of their past, present and future processes, as well as socioeconomic benefits of different types (fishing, cattle farming, recreational use, enjoying nature), in such a way that the most representative and singular sectors conform their Natural Patrimony, within which the hydrological profile (surface and underground waters) is highly relevant.

Part of the region was declared a Protected Natural Space (ENP) with the figures of Regional Park, Natural Reserve and Natural Space of Candelario by the Autonomous Regions of Castile and Leon and Extremadura (Fig. 38.1); Natural 2000 Space (SPA, SCI) by the European Community and Biosphere Reserve by UNESCO, which affords a legal protection regime. Nevertheless, the fragility of the hydrological values and some ecosystems could lead to the deterioration of this legacy; a better knowledge of their importance and meaning is therefore necessary as a guarantee of conservation.

In this work, the inventory and mapping of the Hydrological Patrimony are proposed for use as management tools. Educational itineraries are determined on their basis to facilitate understanding of their essential characteristics and encourage their dissemination.

38.2 Methodology

The aim is to specify a catalogue of the Hydrological Patrimony to ensure its conservation and facilitate its sustainable use and enjoyment, so that it may be preserved for future generations. It is understood that this forms part of the

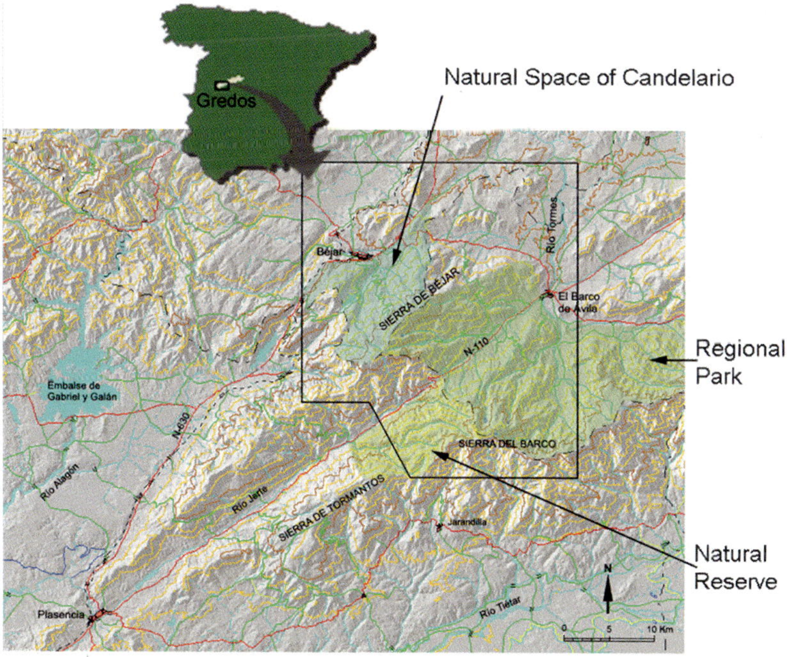

Fig. 38.1 Study area and location of the protected natural spaces

geological heritage, which in turn is part of the Natural Patrimony; therefore, when carrying out the research, the methodology was adapted to those used in these more global studies, so that they would have a connection with each other.

- In the context of World Heritage (UNESCO 1972) three key selection criteria are used: representivity, singularity and beauty, which are taken into account here.
- By Geological Patrimony is understood: "... Set of geologic resources of scientific, cultural and/or educational value, which allow knowledge and interpretation of the origin and evolution of the Earth and the processes that have modelled it, climates and landscapes of the past/present and the source and evolution of life on Earth" (Law 42/2007, of 13th December, on Natural Patrimony and Biodiversity.).
- The GEOSITES Programme (Wimbledon et al. 2000) designed a methodology to identify what the geology of a country represents within its regional frame, through a series of stages and taking into account a series of general considerations and principles, which are assumed.

In this work, Hydrological Patrimony is understood: "The set of hydric processes and forms, where water and its effects demonstrate scientific, cultural and historical singularity", comprising of areas/sites and points of hydrological interest (A/L/PGI_H). The areas represent sectors of great extension with abundant PGI and the Sites, sectors with PGI, which follow a pattern.

Taking into account the proposal by Duran et al. (1998) on Hydrogeological Patrimony: "All those sites in which the water resource presents special interest due to its natural beauty, from an historical point of view, for its importance or use at a given time, for its architectural beauty, directly related to the use of the water, as well as for the volume it represents within the sum of the resources of the community", the different areas/sites and points of hydrological interest (A/L/PGI_H), related to the waters of the Sierra de Béjar and El Barco, in the following typologies were classified as:

1. Singular *geomorphological* elements (processes and forms) of scientific interest, representative of the different morphogenetic systems that have acted in the region throughout the geologic time scale.
2. Singular *ecological elements*, representative of currently existing natural systems with scientific and cultural interest.
3. Singular elements with high *aesthetic quality*, which represent identifying features of the region.
4. Singular elements of *cultural interest*, representative of engineering works related with water use.

Utilizing the Geosites Program methodology, adapting it to the specific regional and thematic environment, a set of 35 A/L/PGI_H, is proposed, selected as records of regional history which stand out for their natural (representivity and singularity), cultural value (cultural manifestations) or their aesthetic beauty (landscapes in which the water is protagonist and distinguishes the site).

38.2.1 Essential Features of the Region, Most Relevant Events and Selection of A/L/PGI_H

To research the Hydrological Patrimony, the characteristic traits and essential features of the region were identified, in terms of age and subject matter within the peninsular unit to which it belongs, from the existing documentation (geological and geomorphological maps, MAGNA Plan, legal norms, doctoral theses and publications in journals) and fieldwork. Its location in the Central Iberian Zone and its geoenvironmental evolution determine that there are different *key contexts* represented with examples, more or less representative of regional history, within the peninsular context.

The region has undergone different stages, the Quaternary being the most significant in hydrological terms. After constitution as relief in blocks and the tilting towards the SW at the end of the reactivation period, the fluvio-torrential incision/engagement began, taking advantage of the fracture zones (late Hercynian and alpine) (Mapas Geológicos de España (553, 576, 599, 577)). Quaternary neotectonics originated new fracturing and tilting, causing reconstructions of the relief, deviations, falls and captures in the hydric network (Mapas Geológicos de España (op. cit)). The climatic oscillations, with cold climates giving way to other warm ones, are a determining factor in the configuration of singular modelling,

Table 38.1 Table of valuation with the A/L/PGI_H more valued of the Béjar and El Barco massifs

A/L/PIG	Criteria of selection and assessment										Interest			
Type	Designation	Representivity	Singularity	Geodiversity	Complexity	Temporality	Beauty	Cultural manifestation	Accessibility	Fragility	Science	Didactic	Tourist-C	Value
G	Lagons: Duque/ Trampal	3	3	3	3	3	3	2	3	3	10 (9)	9.5 (8.5)	6.5 (8.4)	26
G	Giant'sPotholes Los Infiernos	3	3	3	2	3	3	3	3	3	9.5 (8.5)	9.5 (8.5)	7 (9)	26
G	Waterfalls in Bonal Gorge	3	3	2	3	3	3	2	3	3	10 (9)	9 (8.1)	8 (9)	25
G	Lagoon/Cirque Caballeros	3	3	3	3	3	3	1	1	3	10 (9)	8 (7.2)	5 (6.4)	23
G	Lagoons La Nava	3	3	2	3	3	3	2	1	3	10 (9)	8 (7.2)	5 (6.4)	23
C	Landscape Peña Negra	3	2	2	3	2	3	3	1	3	7 (6.3)	8.5 (7.65)	5.5 (7.07)	22
G	Waterfall of Las Yeguas	3	2	2	1	3	3	1	3	3	8 (7.2)	7.5 (6.75)	5.5 (7)	21
E	Peat Bogs of Tremedal-PN	2	3	2	2	2	2	2	3	3	8.16 (7.3)	7.16 (6.4)	5.16 (6.5)	21
G	Mat-grasslands Venerofrio	2	2	2	2	2	2	3	3	3	6.66 (5.9)	7 (6.3)	5.66 (7.2)	20
G	Mat-grasslands Barco Gorge	2	2	1	2	2	3	3	3	3	6.66 (5.9)	7.16 (6.4)	5.16 (5.6)	19
G	Stone circles of Nava Gorge	3	3	2	1	2	2	-	1	2	7.66 (6.8)	5.16 (4.64)	3.16 (2.84)	16

Fig. 38.2 PGI_H with highest score (in order corresponding to Table 38.1) on the basis of the inventory, a *mapping of* Natural Patrimony with *A/L/PIH$_G$, A/L/PIGeoenvironmental and* quality of landscapes was drawn up for Spatial Planning of the PNS (Fig. 38.3; Cruz et al. 2008)

evidence of these climatic periods (Mapas Geológicos de España (op. cit)). In the current tectonic and climatic frame, consolidated after the definitive withdrawal of the ice, different natural systems with high singularity and fragility were developed. The wetland areas with their abundant endemisms cover a certain extension and are of great importance, not only scientific, but also economic and social.

Once the essential features are identified, the basic contexts are detailed, related to the typologies established and the relevant events and occurrences specified, considering characteristics that pinpoint key times of the phenomenon, singular qualities and features, at least at regional level.

Selection of the A/L/PGI_H was carried out after evaluating examples referring to the different elements of interest, obtained from the existing documentation and fieldwork, considering their *representivity, singularity, geodiversity, complexity, cultural manifestation, and temporality covered, beauty and fragility* (Table 38.1; Figs. 38.2 and 38.3). Finally, an Inventory was drawn up with the A/L/PGI_H that presented special features owing to their rarity, cultural manifestation, beauty or history, so that they form part of the Natural-Hydrological Patrimony of the region (Table 38.2).

38.2.2 Update: Didactic Divulgation of Patrimony

The inventory and mapping obtained serve as the starting point for the selection of areas in which to make nine itineraries, which enable a sufficient number of PGI_H, $PGI_{M/G}$, PGI_P, PGI_A (Points of Geomorphological/Geological, Landscape and Environmental Interest) to be seen (Cruz et al. 2010).

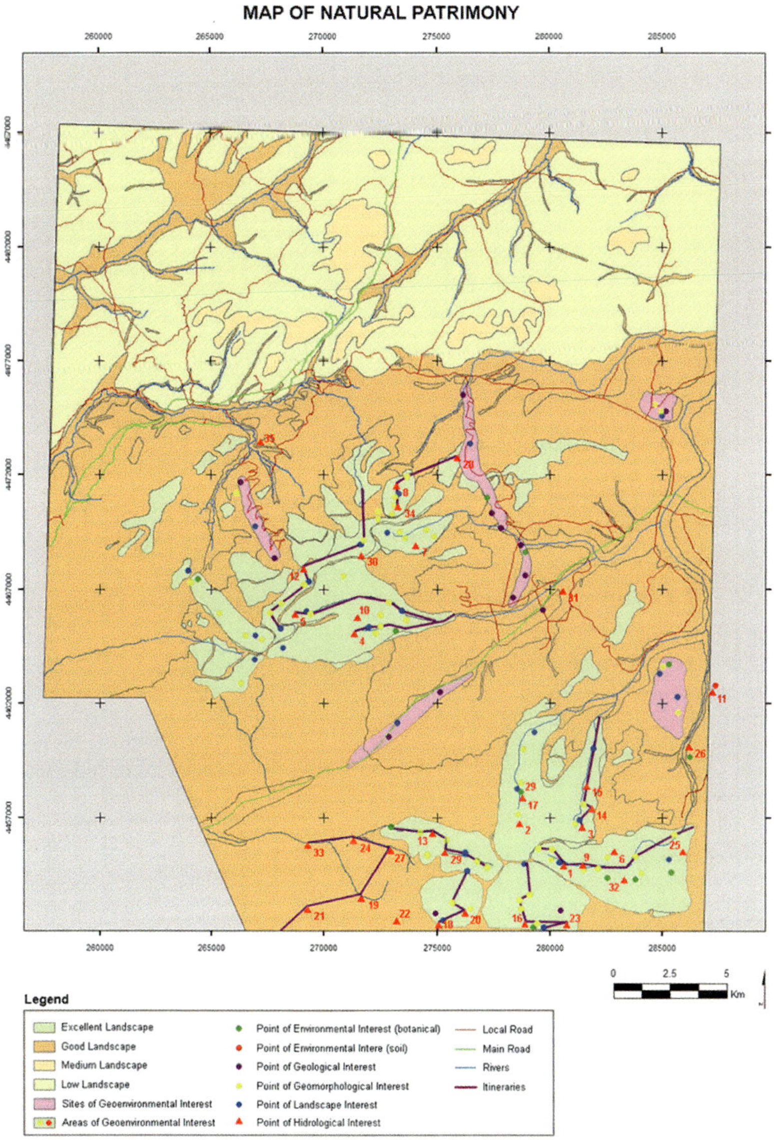

Fig. 38.3 Natural patrimony map of the Béjar and El Barco massifs (PGI_H, PGI_G, PGI_M, PGI_P, PGI_A, PGI_E

These two tools (inventory and mapping) may serve as suitable instruments for the preparation of many other training and communication actions concerning the patrimony, such as posters, information panels, informative leaflets, routes, etc. All of this will undoubtedly result in the greater knowledge of these values and thereby in their greater conservation.

Table 38.2 Inventory of A/L/PGI_H of the Béjar and El Barco massifs

Type		N°	A/l/PGI	Area/Site/Hydrogeological point of interest		Interest/Value
Geomorphological	**Glacial**	**1–5**	**A**	**Glacial Cirques/ Lagoons**	Lagoons: Caballeros, Barco, Nava , Duque and Trampal (3)	
		6–8		**Morrenas sequence**	Morrenas de Caballeros, El Endrinal and Peña Negra	Glacial representivity (Climate sequences)
		9–10	**L**	**Forms fluvioglacial**	P forms in Gorges of Caballeros and El Duque	Singularity (Glacial Epoch)
	Periglacial	**11**	**P**	**Periglacial deposits**	Gelifluidal cone of Navalonguilla	Singularity (Ancient periglacial)
		12			Field of Blocks in Cuerda del Calvitero	Singularity post-Montera periglacial
		13–14	**L**		Periglacial scree in Valle and Circo de Serrada and La Nava	Representivity(Climate sequences) Periglacial singularity
		15		**Periglacial soils**	Stone circles of Nava Gorge	
		16	**P**		Hydro laccoliths of Cuartos Gorge	Representivity(Active periglacial process)
		17			Cushioned soils in Barco/Serrada Gorges	Periglacial singularity
	Fluviotorrential	**18**	**A**	**Fluviotorrential valley**	Jaranda Valley	Representivity Aesthetic beauty
		19–20	**P**	**Waterfalls**	Waterfall of Las Yeguas and Jaranda	Fall sequence singularity 20 – 40 m
		21–23	**L**		Waterfall Sequences in Bonal Gorge (3), Mayor (2)and Cuartos (3)	Aesthetic beauty
		24		**Rapids**	Rapids and Potholes of Asperones Gorge	Representivity: Falls – Singularity Seq. rapids 0,5–1 m
		25		**Alluvial cones**	Cone sequences in Caballeros	Representivity
		26		**Alluvial terraces**	Terrace sequences in Caballeros	Climate sequences
Ecological	**Wetland zones**	**27–28**	**L**	**Peat bogs**	Peat Bogs of Panera and Tremedal-Black P with glacial endemisms	Fragile ecosystem Representivity

		29–30		**Mat-grasslands**	Mat-grasslands in Barco/ Serrada Venerofrio Gorges	Glacial endemism
		31	A	**Edaphic-climatic forests**	Aravalle riverside woodland	Representivity
		32	L		Birch forests in Caballeros Gorge	Fragile ecosystem Aesthetic beauty
Cultural	**Landscape**	**33**	L	**Fluviotorrential landscape**	Rapids and Giant's Potholes in Los Infiernos Gorge..	Representivity Neotectonic forms/falls/potholes Aesthetic quality
		34	A	**Landscape alpine**	Peña Negra Becedas.	Recreational and sporting use
	O.Ing.	**35**	P	**Snow well 16th–18th century**	Snow Well of La Garganta	Representivity Cultural manifestation

38.3 Results and Conclusions

In this work, objective information on the hydrological patrimony was gathered for possible use as a spatial planning tool for the territory:

- An inventory of the A/L/PGI_H was carried out, classified and evaluated, with the most significant hydrological aspects of the region; important information was provided on morphology, hydric processes and paleoenvironmental events that took place in the Quaternary period.
- The mapping of the Hydrological Patrimony was drafted and computerized by GIS for application as a useful tool in the management of these natural spaces.
- Eight geoenvironmental itineraries were designed and drawn up through PGI_H and $PGI_{M/G/P/A}$ and one hydrological PGI_H to be used in education and communication.
- The inventory obtained may form part of the future Spanish inventory, anticipated in Law 42/2007 on Natural Patrimony and Biodiversity (BOE 299/2007).
- The patrimonial wealth makes it possible that this region, along with the adjacent ones, which also have their own rich heritage, may be included in the UNESCO Geoparks Program, so that their patrimonial set goes on to play an active role in the sustainable economic development of this region.

Acknowledgments The authors are thankful for the support from the Projects SA041A08 (JCyL) and CGL 2008-03998 and 04000/BTE.

References

Cruz Ramos R, Goy y Goy JL, Zazo Cardeña C, Martínez Graña A (2008) Cartografías Geomorfológicos y paisajísticas del Patrimonio Natural para la Gestión de los Espacios Naturales Protegidos de Candelario y Gredos (Salamanca-Ávila, España). En: Benavente J, Gracia FJ (eds). Trabajos de Geomorfología en España. SGE, 439–442

Cruz Ramos R, Gaite Cuesta M, Goy y Goy JL, Zazo Cardeña C (2010) Importancia del Patrimonio natural y paisajístico en los Espacios Naturales del Sur de Salamanca y Norte de Cáceres. Puesta en valor mediante itinerarios geoambientales interactivos. En: Una visión multidisciplinar del patrimonio geológico y minero. Cuadernos del Museo Geominero, n° 12. 373–387

Duran JJ, Vallejo M, Fernández L (1998) Patrimonio Hídrico e hidrogeológico de la Comunidad Autónoma de Madrid. In: Durán JJ (ed) Patrimonio Geológico de la Comunidad Autónoma de Madrid. Sociedad Geológica de España, Madrid, pp 193–217

Mapas Geológicos de España, escala 1/50.000 (Segunda Serie) Hojas: Béjar (553), del Valle C (576), de la Vera C (599), Bohoyo (577). Plan Magna. Ed. Servicio de Publicaciones de I.T.G. E. Madrid

Wimbledon WAP, Ishchenko AA, Gerasimenko NP, Karis LO, Suominen V, Johansson CE, Freden C (2000) Geosites an IUGS initiative: science supported by conservation. In: Barettino D, Wimbledon WAP, Gallego E (eds) Geological heritage: its conservation and management. Instituto Tecnológico Geominero de España, Madrid (Spain), pp 69–94

Chapter 39
Observations About the Geology of Western Cuba, Most Significant Aquifers, and the Geomorphology of its Catching Landforms

R. Gutiérrez Domech

Abstract The Republic of Cuba has a high degree of geological knowledge of its territory. In the western part (from Pinar del Río, in the west to Matanzas, in the middle west) there are some of the more significant aquifers of the Cuban archipelago. This is a clear karst territory, but the study of geological characteristics of the rocks, and the relationship among these and the physic–mechanic properties and the karst processes are nowadays insufficient to evaluate the speed of the processes, the probable geological risks and the karst environment to the environmental aggressions. The Jaimanitas and Vedado geological formations that constituted the base of the karstification at the Guanahacabibes peninsula, the western edge, contain the underground waters of this region, even though the flows are very limited. Guasasa, Jagua, Artemisa, Francisco, Sierra Azul (Pinalilla) among others, are the geological formations that contain waters in karst fissure aquifers at the Guaniguanico mountain range, Pinar del Río province axis. This also presents big pool remnants of the river and stream currents that cross the calcareous hills. Paso real formation constitutes the main stratigraphic unit that is water bearing at Pinar del Río southern plain. At Isla de la Juventud there are also small aquifers associated with karst landforms developed on Colombo, Playa Bibijagua, Sierra de Caballos and Sierra Chiquita marble to the north and the Jaimanitas, Cocodrilo and Vedado limestone formations to the south. The aquifers in the protected areas require proper attention considering the special conditions of the locality.

Keywords Karst • Geological formations • Cuban archipelago

R. Gutiérrez Domech (✉)
Instituto de Geología y Paleontología, La Habana, Cuba
e-mail: rgdomech@igp.gms.minbas.cu

H. Farfán González et al. (eds.), *Management of Water Resources in Protected Areas*, Environmental Earth Sciences, DOI 10.1007/978-3-642-16330-2_39,

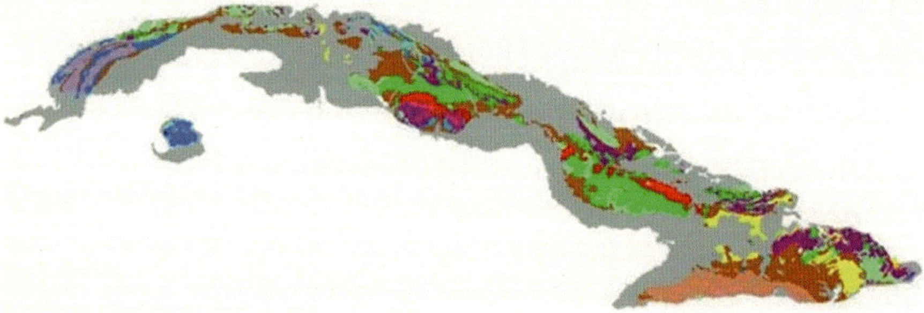

Fig. 39.1 Cuban Republic geology map. Western edge is Guanahacabibes peninsula with Pliocene-Pleistocene covering rocks as well of the southern plain. Axis of Pinar del Río is Guaniguanico mountain range with Jurassic – Cretaceous rocks, East are La Habana, Ciudad de la Habana and Matanzas, provinces with more area of covering Quaternary and Neogene's layers. The smaller island south is Isla de la Juventud. All this territory is western Cuba

39.1 Introduction

Cuba is an archipelago where the Island of Cuba (105,007 km^2), Isla de la Juventud (Isle of Youth) named Isla de Pinos before (2,200 km^2), Cayo Romano (926 km^2), Cayo Coco (370 km^2) and 4,192 other small islands and keys, make a surface of more than 110,000 km^2 (Fig. 39.1).

About 44,000 km^2 of the Cuban territory shows naked karst ore cover by thin soil layers at mountain and plain relief.

Close to 18,000 km^2 are covered by thicker soils in hilly and plain relief, occupying the bottom of some rich valleys, and nearly 8,700 km^2 are buried below very thick soils of various origins and characteristics at flat zones. On these regions are 151 of the 165 recognized aquifers of the country. From these underground waters is based the water supply to the largest cities and more important agriculture plans and industries (Gutiérrez 2003).

Many of these aquifers are partially or totally on protected zones; and, as a part of it, most receive a proper treatment that preserves its quality and guarantees efficient use.

39.2 Geological Formations Where the Aquifers Are

Guasasa Formation is located at Sierra de los Organos, Pinar del Río province, with an Upper Jurassic (Upper Oxfordian) to Lower Cretaceous (Valanginian) (Díaz et al. 2001). These are micrítica limestones, calcareous sandstones, and flint lens. Limestones are sometimes granular and frequently laminate in few members; however, in San Vicente member (Kimmeridgian), the one with the larger exposition and thickness limestones are massive with dark colors.

Compression resistance (Rc) of these rocks is from 815 to 1,000 kg/cm^2 at the massive but lower on the laminate ones (Gutiérrez and Barrientos 2003).

Transmissibility depends on the cracking degree and the karst passages, because porosity is almost none. However, there are heavy flow springs like those in Mal Paso, Sierra de Mesa basin, and Los Portales, with mineral sulphate, sulphide calcium waters.

Karst development is strong on the surface and underground, and at the contact with the Jagua Formation some of the largest Cuban caves have been excavated, such as Palmarito cave system (more than 50 km), Gran Caverna de Santo Tomas, Majaguas-Canteras cave system, Cueva Fuentes and others, which were caved by rivers coming from the surrounding shale hills that go across, following karst passages, the karst towers (mogotes), as well receive the rain water infiltration from the top of the mogotes (Gutiérrez and Iturralde-Vinent 1990).

Artemisa formation is distributed at Sierra del Rosario and NE of the Alturas de Pizarras del Norte, Pinar del Río province, which has Upper Jurassic (Middle Oxfordian) and Lower Cretaceous (Valanginian).

Micritic limestone, well bedded in thin to medium layers, calcilutitas, calcareous sandstone and radiolarian silician insert are the main rocks. On the base sometimes there appear dark colors of limolitas and thin grain sandstones (Díaz et al 2001).

The Rc is around 815 kg/cm^2 because of the high cracking degree and bedding. This parameter can be smaller to 600 kg/cm^2 in saturation conditions (Oliva 1984).

The springs and aquifers of different flows are many; however, the confinement that produces the common insertion of impermeable layers with the karst limestone, lead to a very fast underwater flow and quick discharges.

This formation also presents springs like those of San Diego de los Baños, with calcium sulphate and sulphite waters or those on Soroa with calcium-bicarbonated-magnesian and sulphite waters.

Many of the rivers that go across Sierra del Rosario, where this formation is leading, show partial underground courses because of caving and deep river canyons.

Surface landforms are abundant and of different types. The largest caves are Los Perdidos, del Altar and Santa Cruz river system.

Francisco formation covers the SW of Sierra del Rosario, with clays, silts, micrítìc limestone, schist and insert sandstone, sometimes with calcareous concretions into schist and limestone. Limestones are frequently laminated. The Age is Upper Jurassic (Middle Oxfordian high part – Upper Oxfordian, low part) (Díaz et al. 2001).

The Rc is around 815 kg/cm^2, being lower with saturation and bedding (Gutierrez and Barrientos 2003).

This formation presents surface karst landforms that are not well developed, because the structural characteristics of the rock pack are frequently fragmentized. Aquifers are between layers, but can have heavy flows because of the transmissibility of the rocks.

Pinalilla (Pan de Azúcar) formation is well developed on the Pan de Azucar massif, at Sierra del Rosario. The Age is Lower Cretaceous (Albian) – Upper Cretaceous (Turonian), and it is made of massive limestone and thick to massive strata, grey to green-grey colour. Locally, very thin layers of lime and clay insert with the limestone can be found. Fragments of bauxite are common on the surface (Gutiérrez et al 2007).

The Rc is from 1,000 to 1,300 kg/cm^2, low porosity, even when the heavy cracking makes it easy for the underground rivers to cross and causes the storage of great amounts of water in cave passage depressions. In these rocks there are many caverns such as Canillas Cave System and others of various shapes and sizes (Gutiérrez and Barrientos 2003).

There are well developed surface karst landforms, such as crack lapiaz (kluftkarren) and wall lapiaz (wandkarren) (del Busto and Gutiérrez 1986).

Paso Real formation is one of the most extensive units in the Cuban territory, as a large non-continuous belt from Guane in Pinar del Río province and S of Holguín province.

The southern Pinar del Río plain constitutes more important province aquifers, locally affected by seawater, because of overexploitation.

Alternations are composed of organogenous biohermic limestone, clayey limestone, dolomite limestone, calcareous sandstone and marls, with the insertion of sandstone and clays (mainly smectite) with gypsum, halite, pyrite and lignite. The Age is Oligocene to Upper Miocene.

Jaimanitas formation is present in a non-continuous belt that covers almost all the Island of Cuban shoreline and near small islands on the archipelago. Because of its stratigraphic position and geomorphology, it is pointed as the Upper Pleistocene Age.

Massive biodetritic limestones, generally karstified, with many fossils as shells, corals of actual species and occasionally karst pockets, fill with a very thin mixing of calcareous-clay reddish color. These go to massive calcareous sandstone of well bedded and sometimes insertions of marls. Cementation is variable with white to yellow colour.

The Rc goes from 180 to 500 kg/cm^2, depending on the porosity lower than 10 %, and shows strong development of surface karst landforms contributing to this del marine abrasion and also shows many underground forms such as cueva de los Pájaros, La Tomasa, Las Cenizas and others. Aquifers are non-continuous and almost all are affected with marine intrusion (Gutiérrez 1982).

Vedado formation goes as a narrow and non-continuous belt about the north shore of the La Habana, Ciudad de la Habana and Matanzas provinces and at the south coast of Pinar del Río, Matanzas and Cienfuegos (in the south central) provinces.

The Age is Upper Pliocene to Lower Pleistocene and goes from organic, coralline and seaweed and biodetritic, massive or with unclear bedding locally. Sometimes limestone is very hard, others porcelain type, partially with good porosity and caving, re-crystallized, with corals *Acropora prolifera* in a growing position or fragments of it, frequently with dolomite. Can present calcareous sandstone lens. White, yellow some times soft pink colour.

Rc has been measured from 77.35 to 550.19 kg/cm^2 dry situation and from 181.47 to 444 saturated. The porosity is about 10.8 % (Gutiérrez 1977–1979, 1982).

There are many karst landforms like lapies, sinkhole and cenotes, on the surface and caves such as La Santa, la Virgen, Caupolicán, and others of the provinces of La Habana and Ciudad de la Habana (Gutiérrez and Acevedo 1985). Vedado formation outcrops and covers part of the Cienaga de Zapata, where the deepest borehole cutting of these layers had been dug.

Also Vedado formation is the more important base of the famous cenotes of this region.

Güines formation is well developed at the provinces of Pinar del Río, La Habana, Ciudad de la Habana, Matanzas, Cienfuegos, Villa Clara, Sancti Spíritus and Ciego de Avila.

It is compounded by biodetritic limestone fine to medium grain, rich in fossils, organic limestone, dolomite limestone, dolomite and calcareous sandstone. Dolomitization is secondary, generally massive, and rarely bedded (Gutiérrez et al 1995).

Specific weight (Gs) has been measured from 2.63 to 2.77 g/cm^3. It presents porosity from 7.0 % to 38.2 %. The Rc goes from 1,123 to 797 kg/cm^2 but if the rocks get recrystallized, they can reach 1,500 kg/cm^2. In a saturated way the values goes from 90 to 548 kg/cm^2 (Gutiérrez and Acevedo 1985b).

According to Rodríguez (1996) the medium electric resistivity is about 139.8 Ω.

In this formation, rocks are some of the more important aquifers of Cuba, such as Cuenca (Basin) Sur, Cuenca de Vento and also cuencas Almendares-San Juan and partially, Jaruco-Aguacate (Iturralde 1982).

There is a great karst development, surface landforms as dolines, several types of lapiaz and also caves like Boca de Jaruco cave system in the north; cueva García Robiou, de Loma Candela, del Coronel, in the south-central area of La Habana province and cuevas de Torrens, La Torre, Insunsa, El Túnel, and many more in the southern plain of this territory.

Jaruco formation is developed in patches in irregular ways on the north and central part of La Habana province and in the western part of Matanzas province. It has biodetritic limestone, sometimes clayey, frequently with large forms inserted with marls.

There are also reef limestones, calcareous sandstone and calcareous gravels.

Records of Rc saturated had been measured from 105 to 524 lg/cm^2 and in dry tests from 451 to 797 kg/cm^2. Gs goes from 260 to 277 g/cm^3 and resistivity is about 443 Ω (Gutiérrez and Acevedo 1985).

Almendares-San Juan and Jaruco-Aguacate underground basins are partially constituted by these formation rocks and boreholes yield flows from 70 to 200 m/day.

Karst can be sharp at the surface, but corrosion is aggressive in underground caving cuevas del General Aguirre, Sitio Perdido and el Mamey, among others at Escaleras de Jaruco mogote hills.

Cojimar formation is reported from the provinces of La Habana, City of la Habana, Matanzas and PInar del Río, with marls, calcareous marls, clayey marls, calcareous sandstone with a marl matrix and clays.

Rc is low among 50 to 173 kg/cm^2 porosity, which goes from 4 % to 18 % and presents a medium electric resistivity of 118 Ω (Gutiérrez and Barrientos 2003).

Surface landforms are almost absent but there are small caves in several localities where limestone is less clayey and in the geological contact with Güines formations that is above (Gutiérrez 1969–1975, 1980–1984).

Age is Lower Miocene to Middle Miocene.

Canimar formation develops as an irregular and narrow belt from the vicinity of Bacunayagua bridge (at the Habana – Matanzas border) on the western to the eastern section of the Canimar river section, E of Matanzas city. It has organic limestone, clayey limestone with insertions of calcareous sandstone, conglomeratic marls and clays lens.

Gs are from 2.50 to 2.65 g/cm^3. Porosity has values from 10.8 % and 18 %. Rc is low, from 120.9 to 435.3 kg/cm^2 (Gutiérrez and Barrientos 2003).

There are surface landforms but they are not aggressive and also underground passages and chambers as Cueva del Indio, in Yumurí river canyon and others.

Known water bearing is confined and with bicarbonates calcic waters.

Age is Upper Pliocene to Lower Pleistocene.

Bellamar formation outcrops in the northern part of Matanzas province and in a subsurface way at Ciénaga de Zapata western zone.

Shows sandy marls going to clayey and sandy limestone and calcareous sandstone with marly matrix, sometimes with clay pockets inside the marls.

Rc is from 154.3 to 414.5 kg/cm^2. Karst landforms are very aggressive and the underground ones have a wonderful example with the Bellamar cave system, that include Bellamar, El Jarrito, Gato Jíbaro and Rodriguez de la Fuente cave, in the surroundings and even inside the Matanzas city, and those at Ciénaga de Zapata (Balado 2007).

Age is Upper Miocene – Lower Pliocene.

39.2.1 Isla de la Juventud

In this insular territory the northern part is compounded by metamorphic rocks; there are Colombo, Playa Bibijagua, Sierra de Caballos and Sierra Chiquita formations and Sierra de las Casas is not a formal unit, which is a variety of marbles, dolomite ones, sucrose others dark colours, granitic, with fossil remains in one unit, also fetid, light blue with bands and another light colour with metaflint, dolomite, cuarcite, etc (Franco and de la Torre 1980).

These marbles have an Rc from 1,280 to 1,700 or more kg/cm^2 and very low porosity (0.13–0.25 %) (Krynine and Judd 1961).

Age is recognized as Upper Jurassic in the case of Colombo formation where ammonite impressions were found and Upper Jurassic-Lower Cretaceous on the others.

Marbles show very little lapiaz, but karstic canyons and dolines on the surface and caves, such as Furnia (Shaft) Primero de Mayo, al sierra de Caballos or cueva del Agua and others at sierra de las Casas. The underwater accumulation is much better associated with faults and other cracks than to the existence of true aquifers.

The more productive water bearing in this zone comes from sand and nonconsolidated sediments.

The southern part of the island is composed of rocks of the Quaternary covering and presents calcareous white and locally light pink sandstone, but with weathering takes on a grey colour. These rocks are well bedded at Cocodrilo formations with very little underground waters. The best water bearing is in Jaimanitas formation layers; this place is composed of organogenic limestone, marly and calcareous sandstone of tick grains, which some authors (Franco and de la Torre 1980) named Punta del Este formation where karst landforms are present as well as cenotes and caves.

Also Siguanea formation, with quartz sands and gravels, can locally contain underground water.

39.3 Conclusions

- The stratigraphic units composed of carbonate rocks constitute the more important aquifers of the western Cuban territory.
- The relationship among rocks physic-mechanic properties, chemical characteristics, rock structure and karst corrosion degree has not been well established.
- Rc is less with humidity, with karst corrosion and porosity.
- The more productive aquifers are those with rocks with medium Rc, which have larger cracking and porosity.
- Neogene geological formations are the more productive on the underground water catching.

References

Balado Evelio J (2007) El karst de la Ciénaga de Zapata. Algunas consideraciones acerca de su génesis y características distintivas. Memorias II Convención de Ciencias de la Tierra. VII Congreso de Geología, ISBN 978-959-7117-11-2

del Busto R, Gutiérrez R (1986) Pequeña Enciclopedia trilingüe de términos geomorfológicos. Editorial Científico Técnica, La Habana, p 224pp

Díaz Otero C, Furrazola G, García D, Linares E, Gutiérrez R et al (2001) Léxico Estratigráfico de Cuba. Instituto de Geología y Paleontología (inédito) 355pp. La Habana

Franco G, de la Torre A (1980) Los depósitos costeros de la Isla de la Juventud (Isla de Pinos), Cuba. Cienc. de la Tierra y del Esp. Academia de Ciencias. La Habana, (2):3–13

Gutiérrez R (1969–1975) Estudios estratigráficos para los levantamientos geológicos de los embalses proyectados: Maurín, Niña Bonita, Cuatro Caminos-Falcón, Ejército Rebelde, La Coronela y otros. Informe técnico. Técnico. Grupo Hidráulico del DAP

Gutiérrez R (1977–1979) Estudio ingeniero geológico e ingeniero geomorfológico para el proyecto técnico de la Central Termoeléctrica del Este de la Habana (CTE de Santa Cruz). Informe técnico, MICONS

Gutiérrez R (1980–1984) Estudio ingeniero geológico e ingeniero geomorfológico para el proyecto técnico-ejecutivo del Centro Nacional para la Producción de Animales de Laboratorio (CENPALAB). Informe técnico. Empresa de Investigaciones Aplicadas No. 2, de la Empresa Nacional de Investigaciones Aplicadas del MICONS

Gutiérrez R (1982) Estudio ingeniero-geológico de una zona cársica en el NW de la provincia de La Habana. Proceedings Coloquio Internacional sobre Hidrogeología cársica en la Región del Caribe

Gutiérrez R (2003) Problemas originados por el manejo inadecuado del medio cársico en lugares de emplazamiento de diversas obras, en Cuba. Memorias V Congreso Cubano de Geología y Minería

Gutiérrez R, Acevedo González M (1985a) El carso litoral y las cuevas de Boca de Jaruco, costa norte de la provincia de La Habana. Libro de Resúmenes Simposio XLV Aniversario de la Sociedad Espeleológica de Cuba

Gutiérrez R, Acevedo González M (1985b) Notas sobre el Sistema Cársico de Tirabeque, Bejucal. Libro de Resúmenes Simposio XLV Aniversario de la Sociedad Espeleológica de Cuba

Gutiérrez R, Barrientos A (2003) Las unidades litoestratigráficas carsificadas de Cuba Occidental. IV Congreso Cubano de Geología y Minería, 2001, ISBN 959-7117-10-X

Gutiérrez R, Iturralde Vinent V (1990) Condiciones geológicas de formación del carso en Cuba. Congreso Internacional L Aniversario de la Sociedad Espeleológica de Cuba

Gutiérrez R, Lavandero R, Martínez Salcedo J (1995) Condiciones geológicas de los yacimientos de minerales en regiones cársicas de Cuba. Memorias LV Aniversario de la Sociedad Espeleológica de Cuba. Primera Reunión Iberoamericana

Gutiérrez R et al (2007) Protección y conservación del Patrimonio Geológico. Provincias Occidentales. Informe técnico. Provincia de La Habana. Instituto de Geología y Paleontología, MINBAS

Iturralde MA (1982) Evaluación ingeniero-geológica del territorio de las provincias de la Habana. En: Contribución a la Geología de las provincias de la Habana y Ciudad de la Habana, pp 136–150. Editorial Científico-Técnica, La Habana

Krynine D, Judd WR (1961) Principios de Geología y Geotecnia. Editorial Omega

Oliva V (1984) Propiedades mecánicas de las rocas. Editorial Oriente, 146pp. Santiago de Cuba

Rodríguez W (1996) Evaluación de las. En: Contribuciones a la hidrogeología y al Medio Ambiente en Cuba. Publicaciones Universitat Jaume I, pp 137–154, España

Chapter 40
The Importance of the Alexander von Humboldt's Work "Island of Cuba" to the Study of Caves in Cuban Territory

L.E.Panisset Travassos

Abstract The author believes that it is not possible to dissociate Alexander von Humboldt (1769–1859) from the physical studies of nature, and when one thinks about Geography, this separation becomes even more difficult to be made. Several authors have considered Humboldt the founder of this science, while others state that the naturalist is responsible for the foundation of the so-called modern science as a whole. Despite the existence of many articles written about Humboldt and his research, some still do not recall his importance in the study of karst and caves, including some studies made on the island of Cuba. Reading the 397 pages of the book, *The Island of Cuba* in English (1856), it is possible to identify references to the karst phenomena, descriptions of regional limestone and its relation to the existence of water sources, for example. In the present work it is possible to say that many of the regions visited by Humboldt are nowadays acknowledged at the National System of Protected Areas of Cuba (SNAP – Sistema Nacional de Áreas Protegidas de Cuba) which consists of the country's most important natural areas, both in land and at sea. Humboldt's travels to the equatorial regions of the world represent an important contribution to the development of Karstology in Latin America and the Caribbean.

Keywords Physical studies of nature • Cuban karst • Alexander von Humboldt

L.E. Panisset Travassos (✉)
Pontifícia Universidade Católica de Minas Gerais – Graduate Program in Geography, Prédio Emaús, Av. Itaú n°505, Bairro Dom Bosco, Belo Horizonte, MG, Brazil
e-mail: luizepanisset@gmail.com; luizpanisset@uol.com.br

H. Farfán González et al. (eds.), *Management of Water Resources in Protected Areas*, Environmental Earth Sciences, DOI 10.1007/978-3-642-16330-2_40,

40.1 Introduction

It is believed that, it is not possible to dissociate the name of Alexander von Humboldt (1769–1859) from the physical studies of nature. If one thinks about Geography, this separation becomes even more difficult. Turley (2001) says that many consider Humboldt the founder of this science, while others see him as the individual who is the most influential individual in modern science as a whole.

Kohlhepp (2006) agrees that Humboldt contributed significantly to the development and consolidation of Geography as science. Working as a physical geographer, he started climatological and phytogeographical studies; as a human geographer, he was engaged both in relevant geopolitical aspects and regional studies of human geography; and, as a cartographer, he represented nature very didactically through maps and beautiful profiles. Humboldt had a major role among the painters of the "exuberant tropical New World". This statement is proven by Diener (2007) that identifies, for example, how his works influenced the paintings of Johan Moritz Rugendas (1802–1858).

Naturalist Carl Friedrich Phillipp von Martius (1794–1868), like many of his time, was also influenced by Humboldt. Henriques (2008) reminds us of this fact when he says that "*as a European, Martius certainly had great curiosity for the Americas, due to reading Humboldt's*", which is often cited in the journals of his travels to Brazil.

Tuan, cited by Rodaway (2007), comments that Humboldt, who is widely known for having explained the physical world, was among the first to use the representation of landscape and poetry to broaden the geographical experience with feeling and emotion.

It is worth emphasizing the fact that "*he knew how to combine the concentration of his scientific activities and enthusiasm in a clear and communicable way*" (Kohlhepp 2006). On examining Humboldt´s works one can say, with no doubt, that he deserves all the "titles" and countless honors awarded.

When embarking his famous expedition in Latin America in 1799, accompanied by Aimé Bonpland, a French botanist, Humboldt already had considerable geographical, chemical, botanical, mineralogical and physical notions. After returning to his country he published several books and his name was raised to a place in the history of science because of this. In the words of Turley (2001), being tireless as an "*explorer and adventurer, an enthusiastic student of other sources of information and a skilled observer and writer, Humboldt redesigned the geographical perception of the world*".

In Europe, he developed the image of the "New World" and then presented "*the European scientific circles, all the facets of the tropics of Central and South America, and their differentiations.*" (Kohlhepp 2006).

Riesco Jr. (2004) states that more than any other scholar of his time, Humboldt influenced science in several countries. He set new ways and methods of investigation, always looking for their inter-relationship to exceed the limited comprehension of isolated phenomena.

Although there are countless scientific papers about Humboldt and his research, many people do not know how important he is to the study of karst and caves. In his *"Personal Narratives"*, *"Views of Nature"*, *"Cosmos"*, *"Essay on the Superposition of Rocks in Both Hemispheres"*, *"Political Essay on the Kingdom of New Spain"* or in *"The Island of Cuba"* he describes caves in massive carbonate and granite outcrops, ponors, and minerals.

40.2 Karst Features in the "Island of Cuba"

Even after more than 150 years after the first publication, the latter work is still up-to-date. When the various aspects of geography and the social sciences are considered, this work is still a key source for studies of the nineteenth century in Cuba.

According to Humboldt (1856), he has been "*twice in Cuba: for three months on one occasion, and for a month and a half on the other*". He continues by stating, "*I have had the good fortune of enjoying the confidence of persons, who could give me reliable information from their talents and position, either as proprietors, administrators, or merchants*" (Humboldt 1856). Faithful to the geographical principle of comparison, Humboldt (1856) said that the "view of Havana from the entrance to the port is one of the most picturesque and pleasing ones on the northern equinoctial shores of America" but that it was not as luxurious as the view of Rio de Janeiro, Brazil.

As the shores of the island of Cuba are covered with cays and reefs through more than two-thirds of their extent, and the navigable channels lie outside of these obstructions, the true figure of the island was for a long time unknown. Its width, particularly between Havana and Batabano, has been exaggerated, and it is only since the Hydrographic bureau at Madrid, the best establishment of its kind in Europe, has published the labors of Capt. Jose del Rio and Lieut. Yentura Barcaiztegui, that its area has been calculated with any degree of accuracy (Humboldt 1856).

The geological setting of the Island is composed by "*secondary and tertiary formations, through which granitic-gneiss, syenite, and euphotide rocks have protruded*" (Humboldt 1856). However, the most important parts concerning some protected areas of Cuba nowadays are the central and western parts which "*contain two formations of compact limestone: one with sandy clay, and the other with gypsum*" (Humboldt 1856). Regarding the first formation, he does not mention the age but makes comparisons with the formation of the Jura, in France, stating that "*it is white or of a light yellow ochre color, brittle, sometimes conchoidal and sometimes smooth (...) similar to the spongy strata presented by the Jurassic limestone at Franconia, near Dondorf, Pegnitz, and Tumbach*" (Humboldt 1856).

Humboldt mentions the limestone "*modern formation in the hills of San Juan, near Trinidad whose peaks remind me of the limestone mountains of Caribe [Venezuela] (...) It also contains great caverns near Matanzas and Jaruco (...).*

This frequency of caverns, in which the rains accumulate and the brooks disappear, sometimes causes great disasters" (Humboldt 1856). It is not clear what kind of disasters he is talking about; the subsidence or the collapse of the soil is relatively common in karst areas thus forming dolines.

Another interesting karst phenomena mentioned in a work regarding water behavior is the probable "*submarine communication between the limestone formations of the shore (...). The fresh water of Cuba rises by hydrostatic pressure through the coral rock of the bays, as it happens in the bay of Jagua, where fountains spring forth in the salt water, and are a resort for the Manati*" (Humboldt 1856).

The water resources are mentioned again when Humboldt mentions the southern side of the island, which "*is most abundant in springs, where, from Jagua to Point Sabina, a distance of forty-six leagues, the country is a continuous swamp. The abundance of water that nitrates through the fissures of the stratified rock is so great, that from the hydrostatic pressure, springs can be found in the sea at some distance from the coast*" (Humboldt 1835).

40.3 Final Considerations

Truthful to the original diversity of geography, which goes far beyond the simple study of the physical characteristics of the territory, the naturalist has dealt with political and social issues which were responsible for affecting the status of the territory in the past as well as its future development. It is also important to mention that a National Park there was named after Alexander von Humboldt (Parque Nacional Alejandro de Humboldt) and is inscribed as a UNESCO World Heritage Site since 2001.

Alexander von Humboldt made important contributions to the fields of earth sciences and, if his descriptions of rock formations and some karst phenomena on the Island of Cuba are considered, even to an incipient Latin American karstology. These are the caves of Matanzas, Trinidad and Jaruco. Some fresh water sources in the ocean were also recognized by the author at the Bay of Jagua.

Comparing Humboldt's information with the National System of Protected Áreas of Cuba (SNAP- *Sistema Nacional de Áreas Protegidas de Cuba* 2002–2010) at present, one can notice that in Matanzas there are about 20 protected areas, divided into 5 categories of use. In Trinidad, Province of Sancti Spiritus, there are 15 protected areas in 6 categories of protection, and in Jaruco, Province of La Havana, there are now 11 protected areas divided in 4 categories of protection.

References

Diener P (2007) The picturesque as an aesthetic category in the art of travelers: notes on JM Rugendas's work. Historia (Santiago), Santiago, v.3, Selected Edition

Henriques RPB (2008) A viagem que revelou a biodiversidade. Ciência Hoje 42(252):24–29

Humboldt AV (1856) The Island of Cuba: a political essay. Translated from the Spanish with notes and preliminary essays by JS Thrasher. New York: Derby & Jackson

Kohlhepp G (2006) Descobertas científicas da Expedição de Alexander von Humboldt na América Espanhola (1799–1804) sob ponto de vista geográfico. Revista de Biologia e Ciências da Terra, 6(1): 260–278, Tradução de Luiz Eduardo Panisset Travassos

Riesco R, Jr (2004) Anzas y vigencia de Alexander Von Humboldt. Revista Geografía Norte Grande, n.32, Santiago/Chile, pp 107–114

Rodaway P (2007) Yi-Fu Tuan. In: Hubbard P, Kitchin R, Valentine G (eds) Key thinkers on space and place, vol 4. Sag0065, London, pp 306–310

SNAP – Sistema Nacional de Áreas Protegidas de Cuba (2002–2010) Available at http://www.snap.cu/news.php

Turley G (2001) Humboldt's Gift: a geographer in profile. Mercartor's World, [s.d.], pp 22–25

Index

H. Farfán González et al. (eds.), *Management of Water Resources in Protected Areas*, Environmental Earth Sciences, DOI 10.1007/978-3-642-16330-2,

H

Printed by Printforce, the Netherlands